DATA FOR NURSES

DATA FOR NURSES

Understanding and Using Data to Optimize Care Delivery in Hospitals and Health Systems

Edited by

MOLLY McNETT

Professor of Clinical Nursing; Assistant Director, Implementation Science Core, The Helene Fuld Health Trust National Institute for Evidence-Based Practice in Nursing & Healthcare, College of Nursing, The Ohio State University, Columbus, OH, United States

Academic Press is an imprint of Elsevier
125 London Wall, London EC2Y 5AS, United Kingdom
525 B Street, Suite 1650, San Diego, CA 92101, United States
50 Hampshire Street, 5th Floor, Cambridge, MA 02139, United States
The Boulevard, Langford Lane, Kidlington, Oxford OX5 1GB, United Kingdom

Library of Congress Cataloging-in-Publication Data
A catalog record for this book is available from the Library of Congress

British Library Cataloguing-in-Publication Data
A catalogue record for this book is available from the British Library

ISBN 978-0-12-816543-0

For information on all Academic Press publications
visit our website at https://www.elsevier.com/books-and-journals

Publisher: Stacy Masucci
Acquisition Editor: Rafael E. Teixeira
Editorial Project Manager: Samantha Allard
Production Project Manager: Maria Bernard
Cover Designer: Christian Bilbow

Typeset by SPi Global, India

Contents

Contributors

India Eaton
Institutional Review Board, The MetroHealth System, Cleveland, OH, United States

Joyce Karl
College of Nursing, The Ohio State University, Columbus, OH, United States

Matthew Kaufmann
The Quality Institute, The MetroHealth System, Cleveland, OH, United States

Melissa Kline
Department of Hospital Administration, The MetroHealth System, Cleveland, OH, United States

Sarah Livesay
Adult Health and Gerontological Nursing, Rush University, College of Nursing, Chicago, IL, United States

Molly McNett
The Helene Fuld Health Trust National Institute for Evidence-Based Practice in Nursing & Healthcare, College of Nursing, The Ohio State University, Columbus, OH, United States

Lorraine Mion
College of Nursing, The Ohio State University, Columbus, OH, United States

Wendy Sarver
Department of Nursing Research, MetroHealth System, Cleveland, OH, United States

Kelly Seabold
Department of Nursing, The MetroHealth System, Cleveland, OH, United States

Mary Zonsius
Adult Health and Gerontological Nursing, Rush University, College of Nursing, Chicago, IL, United States

CHAPTER ONE

Introduction: Why data matter

Molly McNett
The Helene Fuld Health Trust National Institute for Evidence-Based Practice in Nursing & Healthcare, College of Nursing, The Ohio State University, Columbus, OH, United States

Contents

Data and nurses

Healthcare delivery has always been influenced by data. Typically, the scientific method comes to mind when referencing data in healthcare; specifically, data used to generate new knowledge and guide technological, biomedical, and pharmaceutical advances. These advances then impact care delivery in health systems. Unfortunately, it takes up to 15–20 years for this new knowledge to translate into actual practice, and equally as long to evaluate if implementation of these new advances impacts health outcomes and care processes.[1] However, there are other types of data in healthcare that are equally important as research data, and have shaped care delivery and nursing practices for decades. These include quality data, financial data, benchmarks, and nursing sensitive indicators. These types of data have shaped evolution of healthcare since its earliest stages.[2, 3] Regardless of the type or size of a healthcare delivery system, nurses continue to remain a core component of care delivery in almost all settings around the globe. As such, they are in an optimal position to critically analyze these distinct types of data, assess applicability to practice, and evaluate outcomes immediately in real time. Nurses remain key stakeholders of data utilization in all healthcare settings.

Data for Nurses
https://doi.org/10.1016/B978-0-12-816543-0.00001-7

Despite the key role nurses have in integrating data in their daily practice, several reports indicate nurses may lack a solid understanding and application of data utilization in practice, specifically with evidence-based practice, research, and quality improvement activities.[4, 5] Recent estimates on nurse competency for evidence-based practice and integration of data into daily care decisions demonstrate significant deficiencies not only in the United States, but on an international level as well.[4–8] Integration of evidence-based practice, and data in general includes the ability to perform a literature search, understand and interpret findings from research, determine applicability of research findings to practice, understanding the differences between quality improvement and research, and the ability to accurately collect, interpret, and disseminate data.[5]

Information on evidence-based practice has been integrated into most educational nursing programs.[9] Nursing students are taught methods of inquiry and how data can be used to guide nursing practice. However, continued application of this content as nurses begin practicing in the healthcare setting is often scattered or infrequent.[5] There is often substantial variation in how health systems educate their nurses about current evidence-based practices within their organizations, and how ongoing application of this knowledge and data utilization is communicated to practicing nurses.[10–12] Particularly for experienced nurses, many have never received formal training on the diverse types of data, methods for data acquisition, evaluation, interpretation, or application to practice. Nurses must remain knowledgeable not only of new advances in their specialty, but of the increasing ways in which data guide their practice and delivery of care.

Data in healthcare

Care in hospitals is increasingly directed by data. In the United States, the Institute of Medicine (IOM) published an influential report highlighting the tremendous gap existing between knowledge generated from research and quality data, and actual care delivery in healthcare systems.[13] As a result, the past decade has been inundated with substantial shifts in care delivery both in the United States (US) and on an international level. Healthcare delivery models have integrated numerous regulatory and accreditation requirements, with mandatory data reporting and alignment with preestablished benchmarks as central core components.

Nurses are not immune to these shifts in care delivery. As the largest component of any hospital workforce, and present at all points-of-care delivery, nurses play an integral role in impacting quality outcomes and delivery of care based on evidence and best practices.[14, 15] In order to have a solid understanding why data matter, nurses need to appreciate the historical development of data in healthcare, and specifically their role in these developments, in order to have an appreciation for the current landscape.

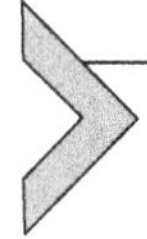

Historical evolution of healthcare and the role of data

Early beginnings: Florence Nightingale

Nurses were at forefront of using data to advocate for change to improve safety and efficiency within hospitals.[16, 17] As early as the mid-1800s, Florence Nightingale emerged as a nursing leader during her tenure in the Crimean War from 1854 to 1856; however, her career after the war focused on healthcare reform, specifically meticulous record keeping providing evidence and rationale for improvements in sanitary conditions for care of soldiers. Florence had utilized data and figures she had accumulated during the war to demonstrate the primary predominance of disease was the main cause of death during the war, rather than actual war injuries.[16, 17] Integrating statistical models, she used these data to make projections about continued poor outcomes among soldiers unless radial changes were made to improve their living conditions. She also used the data as rationale to establish basic standards of hygiene in all hospitals, including structural changes to improve ventilation, contain waste disposal, ensuring access to clean water sources, and improving disinfection practices for bedding and equipment. She encouraged hospitals to keep track of these changes, and record observations with compliance, current practices, and associated outcomes. She remained an advocate for improved hygiene practices, natural ventilation, and access to natural light for hospitalized patients, much of which remain priority agents even today in current design concepts. Florence Nightingale's early efforts to improve standards of healthcare delivery using data provided a substantial foundation for continued healthcare growth.[16, 17]

Advances in the early 1900s

In the early 1900s, a British physician built upon initial work by Nightingale and began to apply an outcomes-based model to healthcare delivery.[2] Earnest Codman collected data on patient progress throughout hospitalization to identify trends in outcomes. He used these data to aid in future decision making when managing patients with similar maladies. Data collected included information about intervention effectiveness, reasons why a certain procedure was unsuccessful, and potential sources of error that could be corrected with future patients. He then partnered with an American physician colleague, Edward Martin, and the two began to advocate for general standards in hospitals based on these data to evaluate quality of hospital processes and care delivery. They formed an initial group of physicians to formally establish The American College of Surgeons, and proposed five basic standards for all hospitals.[18] These standards included: (1) Identification of a definitive group of hospital staff; (2) Establishment of basic credentials and requirements for hospital staff; (3) Hospital staff adopt rules and regulations, including regular meetings to evaluate department specific data and outcomes at specific intervals; (4) Accurate and complete recording and storage of patient data and medical records; (5) Ensuring availability of diagnostic and therapeutic modalities with adequate supervision.[18] These standards incorporated requirements for data collection and storage, recognizing the importance this type of data had in evaluating quality of care and advancing treatment options.

These established standards and collection of data were then used by review committees comprising fellows from the United States and Canada to evaluate hospital applications for formal accreditation in the newly established Hospital Standardization Program.[2, 18] Initially, few hospitals seeking accreditation met criteria due to not having adequate data on processes and patient outcomes; however, after making substantial improvements based on these standards, the number of accredited hospitals dramatically increased throughout the 1920s.[18] These initial standards and processes became the foundation for The Joint Commission on Accreditation of Hospitals.[18]

Period of growth: The mid-1900s

The 1920s–40s marked a period of substantial growth in healthcare due to scientific data, advances in systems thinking, and industrial growth. These developments had an important impact on care processes, equipment, and treatment modalities. Key developments during this time included drug development, such as penicillin and other antibiotics, early data on cancer incidence and establishment of the National Cancer Institute, a better understanding of hormonal regulation and role and development of nutrition and vitamin supplements, and the beginning of regulatory oversight for drugs and devices.[19, 20]

The industrial age not only provided for advancements in equipment, technology, and supplies in hospitals, but processes used in these manufacturing settings were evaluated for efficiency and quality. Once again, evaluation of data yielded valuable information about process metrics, which was then applied in hospital settings.[3, 21] Frederick Taylor was an early pioneer in this field and began to use components of the scientific method to critically evaluate work processes. He made hypotheses about what was causing inefficiencies, gathered data, and evaluated the impact certain process interventions had on work productivity. This approach served as an early foundation in systems analysis and change models, which would later be applied in hospital settings.[21] Similarly, work by Deming also used a systems-level approach to evaluate and reduce variation in work processes, thus increasing efficiency and outcomes.[2, 3] Deming introduced the concept of statistical quality control, and first tested these approaches in Japan after World War II in the 1930s–40s in efforts to improve manufacturing and economic stability.[3] Additional advances included introduction of the Plan-Do-Study-Act (PDSA) approach, which allowed for rapid integration of interventions aimed at root causes or variations, evaluating impact on outcomes in real time, and for additional adjustments of changes to further improve outcomes over time.[3] Advances in these approaches had important impacts in the manufacturing and management realms, and soon became important adjuncts to traditional scientific methods previously used in healthcare models.

In the 1960s, work by Donabedian began to focus on the quality of medical care and evaluation of data to identify methods to optimize care outcomes.[2, 3] He suggested effectiveness in healthcare can best be evaluated through structure, process, and outcomes, and highlighted the importance of evaluating structure of care delivery and its impact on outcomes. His work acknowledged that certainly there were some patient outcomes that couldn't be avoided. However, a critical appraisal of structure and process could help to identify what adverse outcomes could be avoided, and the data were important to critically evaluate to identify errors and adverse outcomes that could be prevented.

In 1965, the Centers for Medicare and Medicaid Services (CMS) was established in the United States as a healthcare delivery model aimed at providing care to the most vulnerable populations—both elderly and economically disadvantaged groups. As a core component of this delivery model, CMS set forth "conditions of participation," which outlined specific components that must be met by healthcare organizations in order to receive financial compensation for care of these patient groups.[2] At the forefront of these conditions was nursing care, along with credentialing systems for providers, and a structure for utilization review that would determine the degree to which these conditions were met. At the time, many hospital utilization review committees were comprised of nurses, due in part to their proximity to care processes, astute observations, and understanding of factors impacting patient care and outcomes. Unfortunately, these initial review committees were only charged with reporting degree of compliance, and not given the authority to propose changes to improve care or meet conditions of participation.[2] Thus, their utility for advancing or improving outcomes based on the data they reviewed was limited.

The premise of these initial utilization review committees was eventually linked to research and academic efforts to begin to offer strategies based on these data aimed at improving outcomes. The National Center for Health Services Research in the United States served at the forefront of these activities and laid the groundwork for Medicare Professional Standards Review Organizations (PSROs), and later Peer Review Organizations (PROs), which focused on decreasing complications and mortality associated with specific disease states and procedures, and decreasing hospital readmission rates.[2]

During this same time period, the initial hospital standards review committees established by The American College of Surgeons had grown to include members from The American College of Physicians, the American Hospital Association, the American Medical Association, and the Canadian Medical Association in efforts to jointly continue to review data on how hospitals were meeting established standards. The work proved exceedingly difficult, as the number of hospitals applying for review continued to grow, at the same time the number of standards required to meet criteria also continued to increase.[18] Thus, in 1951, The Joint Commission on Accreditation of Hospitals was established to become the primary review organization in the United States to evaluate health system compliance with specific standards of care. This model has expanded substantially and remains the primary accreditation body in the United States today.

Evolution and expansion: Late 1900s–2000s

Integration of data evaluation in healthcare systems continued to expand throughout the 1970s into the year 2000. Specific changes included moving from solely utilization review to actual proactive analysis of data in real time to produce expedient outcome improvement, and continuous recording and reporting of data in health systems.[2] In 1970, the National Academies of Science established the IOM, and the Agency for Healthcare Quality and Research (AHRQ). In the United States, both the IOM and AHRQ remain leading agencies driving change in hospitals based on data, and seeking to decrease variations in practice and streamline care delivery. Seminal reports from the IOM specifically highlight not only care deficiencies, but also offer frameworks for potential solutions. These changes shifted a focus from solely utilization review to a critical analysis of data to identify what improves outcomes. Seminal work by Cochrane in the 1970s highlighted the importance of using data from randomized controlled trials to guide care decisions. This was an initial step toward quicker integration of scientific data to further fuel movement toward using data in care decisions, all with the goal of improving quality of care.[2]

These shifts were effective. By the early 1990s, approximately 58% of US hospitals reported having some type of quality improvement program, with an additional 84% of hospitals reporting they had specific plans to integrate a program within the next 2 years.[22, 23] On an international level, surveys of healthcare systems in the US and Canada indicated active quality improvement programs among 88% of healthcare systems by the early 1990s.[22, 23]

In addition to increases in quality improvement initiatives, hospitals were also now using data to guide care decisions. There was a substantial increase in the use of evidence-based practice, which refers to integration of best-available evidence to guide care decisions.[24] Evidence-based practice integrates the best available data from research, clinical expertise, and patient values into care decisions made in the best interest of the patient.[24] Thus, data are at the core of evidence-based practice. Nurses and other healthcare providers were charged with evaluating existing data from research to assess for applicability to current care situations and inform treatment decisions and recommendations. Specifically, within nursing, the field of evidence-based practice rapidly began to expand, with nurses at the forefront of care delivery, and the advent of specific outcomes and quality indicators that were dependent on the quality of nursing care delivered. Now more than ever, nurses needed evidence from research to provide data on which nursing interventions were most effective to decrease falls, development of pressure injuries, ventilator-associated pneumonia, and a myriad of other care interventions.

Recent healthcare changes and the role of data

Perhaps the biggest impetus for healthcare change within recent years lied within the landmark document published by the IOM titled, *To Err is Human*. The document highlighted preventable errors as the cause of almost 100,000 deaths annually in the United States,[25] despite substantial spending toward healthcare, which was up to 18% of the US gross domestic product.[15] A subsequent document in 2001, *Crossing the Quality Chasm*,[13] called upon healthcare providers to use evidence that was now readily available to guide their practices. A chasm between what was known to be effective interventions and what was practiced in hospitals had to be crossed to advance the quality of care. For example, healthcare institutions now had data about the prevalence of hospital-acquired infections and the adverse effects these infections had on patients. Healthcare providers then had an obligation to use this information and identify methods to decrease development of infections for hospitalized patients. Once again, nurses remained key stakeholders in applying this information to their daily practice and holding other members of the healthcare team responsible to optimize outcomes for patients, as a key intervention in infection prevention was hand hygiene.

In the United States, the substantial gap between data and practice in healthcare delivery has been recognized for decades.[13, 26] The advent of the internet in the early 2000s began to change how quickly information was available to those practicing in the field, and began to make important information accessible to the key consumer of healthcare: the patient. As such, the field of informed providers began to grow alongside the increasingly savvy consumer. As such, more attention was shed on why this gap exists, and potential methods to address the gap and bridge data with healthcare delivery were proposed, with numerous agencies at both the national and international levels having a vested interest in applying these data to practice, and using constant surveillance to monitor adherence and adjust as care delivery improved.

In 2003, a core set of competencies was proposed by the IOM for healthcare institutions.[27] These competencies included patient-centered care, interprofessional teams, evidence-based care, integration of quality improvement activities, and use of informatics technology. Development of an institute dedicated to quality and safety education in nursing (QSEN) created specialized curriculum for nurses to meet these specific core competencies, and has now grown to graduate-level programs, and collaborative applications of content with health systems.[9–12] Similarly, development of nurse-led, national institutes for evidence-based practice have emerged and remain dedicated to educating nursing students, and all members of practicing health care teams about the importance of using data to guide care delivery. Development of these resources have become a crucial step toward educating nurses on the importance of data, quality improvement, integration of research findings, systems thinking, and outcomes. Ultimately, nurses must be actively engaged in utilization of data to evaluate the quality and safety of care interventions in real time, and to advocate for change to promote outcomes for patients.[15,28]

Subsequent reports from the IOM specifically identify the need for nurses to broaden their scope to include active participation in research, quality improvement, and patient safety efforts; ultimately they must use data to evaluate effectiveness of care interventions.[13, 14] Inclusion of nursing in these efforts is critical to enhance the quality of care provided, as they represent the largest component of the healthcare workforce.[14] Key recommendations include that nurses practice to the full extent of the education and training, achieve higher levels of education and training, serve as full partners with other members of the healthcare team in shaping care delivery, and integrate better utilization of data for effective workforce planning and policy development.[14] With these changes, a shift toward value of care, rather than volume of services, would be possible, with nursing actively participating and leading collaborative efforts.

In addition to the IOM recommendations, the Centers for Medicare and Medicaid Services (CMS) and other international healthcare organizations have begun to use data to identify and evaluate quality outcomes, which is then used to drive hospital reimbursement and types of services offered. US hospitals have shifted from pay for service models to a focus on quality outcomes, safe patient care, and evidence-based practice, while international health care delivery models have chosen to integrate similar adjustments.[29] A follow-up IOM Report identified 15 core metrics to improve the quality of care in hospitals.[27] Metrics included specific components for evidence-based care and patient safety and quality. On an international scale, the World Health Organization proposed frameworks to assist developing health systems with assessment of resources and existing data to improve quality measurement, reporting, safety, and outcomes.[30,31]

Conclusion

Providing nurses with information about the impact of data on the historical evolutions of healthcare highlights their continued role in data evaluation and application in current care settings. Nurses must possess an understanding of the importance of data, the distinct types of data utilized in healthcare, and critical application and evaluation of data for care interventions. By correctly interpreting, understanding, and reporting quality improvement data, nurses can identify effective interventions in real time and institute changes that improve patient outcomes and organizational reporting. Similarly, astute evaluation of research data by nurses using methods outlined in this book will provide tools for applying findings to practice and methods to evaluate effectiveness. Integration of evidence-based practice, research and quality data, as well as an understanding of relationships between benchmarking and reimbursement will be explored through specific chapters, and applications to clinical practice highlighted by integration of case scenarios. Nurses remain essential leaders and participants for driving and sustaining change, and utilization of data is integral to ensure successful outcomes for patients and health systems.

References

1. AHRQ. *Translating Research Into Practice (TRIP) II. Fact Sheet.* Rockville, MD: Agency for Health Care Research and Quality; 2008. Retrieved from: http://www.ncbi.nlmnih.gov/books/NBK2659.
2. Marjoua Y, Bozic KJ. Brief history of quality movement in US healthcare. *Curr Rev Musculoskelet Med.* 2012;5:265–273.
3. Colton D. Quality improvement in healthcare. *Eval Health Prof.* 2000;23(1):7–42.
4. Pintz C, Zhous Q, McLaughlin MK, Kelly KP, Guzetta CE. National study of nursing research characteristics at Magnet-designated hospitals. *J Nurs Adm.* 2018;48(5):247–258.
5. Melnyk BM, Gallagher-Ford L, Zellefrow C, Tucker S, Thomas B, Sinnott LT. The first U.S. study on nurses' evidenced-based practice competencies indicates major deficits that threaten healthcare quality, safety, and patient outcomes. *Worldviews Evid Based Nurs.* 2018;15(1):16–25.
6. Warren JI, McLaughlin M, Bardsley J, et al. The strengths and challenges of implementing EBP in healthcare systems. *Worldviews Evid Based Nurs.* 2016;13(1):15–24.
7. Saunders H, Vehvilainen-Julkunen K. The state of readiness for evidence-based practice among nurses: an integrative review. *Int J Nurs Stud.* 2016;56:128–140.
8. Bahadori M, Raadabadi M, Ravangard R, Mahaki B. The barriers to the application for the research findings form the nurses' perspective: a case study in a teaching hospital. *J Educ Health Promot.* 2016;23:5–14.
9. Dolansky MA, Schexnayder J, Patrician PA, Sale A. Implementation science: new approaches to integrating quality and safety education for nurses competencies in nursing education. *Nurse Educ.* 2017;42(55):512–517.
10. Koffel C, Burke K, McGuinn K, Miltner RS. Integration of quality and safety education for nurses into practice. *Nurse Educ.* 2017;42(55):S49–S52.
11. Altmiller G, Dolansky MA. Quality and safety education for nurses: looking forward. *Nurse Educ.* 2017;42(55):S1–S2.
12. Johnson J, Drenkard K, Emard E, McGuinn K. Leveraging quality and safety education for nurses to enhance graduate-level nursing education and practice. *Nurse Educ.* 2015;40(6):313–317.
13. Institute of Medicine. Crossing the Quality Chasm: A New Health System for the 21st Century 2001. Retrieved from:http://www.nationalacademies.org/hmd/~/media/

Files/Report%20Files/2001/Crossing-the-Quality-Chasm/Quality%20Chasm%20202001%20%20report%20brief.pdf.
14. Institute of Medicine. The Future of Nursing: Leading Change, Advancing Health 2010. Retrieved from:http://books.nap.edu/openbook.php?record_id=12956&page=R1.
15. Ryan RW, Harris KK, Mattox L, Camp M, Shirey MR. Nursing leader collaboration to drive quality improvement and implementation science. *Nurs Adm Q*. 2015;39(3):229–238.
16. Aravind M, Chung KC. Evidence-based medicine and hospital reform: tracing origins back to Florence Nightingale. *Plast Reconstr Surg*. 2010;125(1):403–409.
17. Kudzma EC. Florence Nightingale and healthcare reform. *Nurs Sci Q*. 2006;19(1):61–64.
18. Roberts JS, Coale JG, Redman RR. A history of the joint commission on accreditation of hospitals. *JAMA*. 1987;258(7):936–940.
19. Pizzi R. Salving with science: the roaring twenties and the great depression. In: *The Pharmaceutical Century, Ten Decades of Drug Discovery*; 2018. Retrieved from:http://www3.uah.es/farmamol/The%20Pharmaceutical%20Century/index.html.
20. Miller J. Antibiotics and isotopes. In: *The Pharmaceutical Century: Ten Decades of Drug Discovery*. 2018 Available from:http://www3.uah.es/farmamol/The%20Pharmaceutical%20Century/index.html.
21. Girdler SJ, Glezos CD, Link TM, Sharan A. The science of quality improvement. *JBJS Rev*. 2016;4(8):e1–e8.
22. Chan YC, Ho SJ. Continuous quality improvement: a survey of American and Canadian healthcare executives. *Hosp Health Serv Adm*. 1997;42(4):5252–5544.
23. Eubanks P. The CEO experience—TQM/CQI. *Hospitals*. 1992;24–36.
24. Sackett DL, Rosenberg WMC, Gray JAM, Haynes RB, Richarrdson WS. Evidence based medicine: what it is and what it isn't. *BMJ*. 1996;312:71.
25. Kohn LT, Corrigan J, Donaldson MS. *To Err Is Human: Building a Safer Health System*. Washington, DC: National Academy Press; 2000.
26. Leach MJ, Tucker B. Current understandings of the research practice gap in nursing: a mixed methods study. *Collegian*. 2018;25:171–179.
27. IOM. *Vital Signs: Core Metrics for Health Care Progress*. The National Academies Press; 2015. Available from: https://www.nap.edu/resource/19402/VitalSigns_RB.pdf.
28. Agency for Healthcare Quality Research (AHRQ). National Guidelines Clearinghouse: Fact Sheet 2018. Available from:https://www.ahrq.gov/research/findings/factsheets/errors-safety/ngc/national-guideline-clearinghouse.html.
29. Centers for Medicaid Services. Patient Protection and Affordable Care Act; Third Party Payment of Qualified Health Plan Premiums. Retrieved from:https://www.federalregister.gov/documents/2014/03/19/2014-06031/patient-protection-and-affordable-care-act-third-party-payment-of-qualified-health-plan-premiums; 2014.
30. World Health Organization. *Everybody's Business: Strengthening Health Systems to Improve Health Outcomes: WHO's Framework for Action*. Switzerland: WHO; 2007.
31. Melnyk B. An urgent call to action for nurse leaders to establish sustained evidence based practice cultures and implement evidence-based interventions to improve healthcare quality. *Worldviews Evid Based Nurs*. 2016;13(1):3–5.

CHAPTER TWO

Quality and benchmarking data in health systems

Kelly Seabold*, Matthew Kaufmann†, Molly McNett‡
*Department of Nursing, The MetroHealth System, Cleveland, OH, United States
†The Quality Institute, The MetroHealth System, Cleveland, OH, United States
‡The Helene Fuld Health Trust National Institute for Evidence-Based Practice in Nursing & Healthcare, College of Nursing, The Ohio State University, Columbus, OH, United States

Contents

Introduction

Understanding data and how they are derived is an essential first step for nurses. It is equally important for nurses to understand how to interpret these data to make meaningful and significant changes to practice within their health systems. A basic step in this process is understanding the use of benchmarking. Gathering and recording serial trended data provides organizations with knowledge of care performance over time; benchmarking then provides an added element of comparison of these data to other health systems, patient populations, and designated care units. Benchmarking is defined as a standardized method for collecting and reporting key performance indicators in a way that allows relevant comparisons among the performances of similar organizations and practice settings.[1] Through the use of benchmarking, nurses are able to determine how unit-specific and

Data for Nurses
https://doi.org/10.1016/B978-0-12-816543-0.00002-9

organizational-level data compared to similar units/organizations, which can then be used to identify areas for improvement. These benchmarks also allow nurses to identify best practices that have led to superior performance.

Identification of benchmarks and relationship to care quality

The use of comparison data in healthcare has been around since the 1600s when medical providers began to explore outcomes such as hospital mortality rates, but it was not until the 1990s that a structured process for healthcare benchmarking emerged with requirements set forth by The Joint Commission.[2] At that time, benchmarking was viewed purely as a measurement tool to identify inequalities among healthcare organizations and rationalize funding in the United States. Healthcare benchmarking has since evolved over time and on an international scale to include a collaborative approach that involves sharing of knowledge, experiences, and best practices to improve care across health systems.[2] This collaborative approach promotes the true purpose of benchmarking in healthcare, which is to improve patient safety, quality of care, and efficiency. As frontline caregivers and the largest, single group of healthcare professionals, nurses have tremendous impact on patient care and outcomes.[3] Given this irrefutable influence on outcomes, nurses must be able to interpret data and benchmarks to elevate nursing practice and improve patient outcomes.

Benchmarking can be established internally or externally. Internal benchmarks are often the initial point of reference for analyzing performance indicators; however, organizations may choose to use internal benchmarking when external benchmarks do not exist or when they wish to evaluate internal performance of key indicators.[4] Internal benchmarks are used primarily in large, multihospital health systems where a significant volume of contributing units can be identified within the system. Common internal benchmarks include organizational mean, specialty-/service-specific means. This form of benchmarking allows for fast reporting and identification of internal problems for standardization of best practices within the health system and is relatively inexpensive to implement.

External benchmarks are often viewed as more beneficial because they provide the added element of comparison against outside organizations, in addition to a larger number of contributing units.[4] Through the use of an external benchmarking vendor, a wide variety of comparison groups or cohorts can be used to produce benchmark statistics for each dataset dependent on the characteristics of contributing organizations, such as metropolitan status, teaching status, or geographic location. These benchmarks can be further classified into unit- or department-specific populations or specialties, such as adult critical care, inpatient pediatric surgical care, or outpatient clinic. Although it is important to compare data against the most similar unit, as the comparison groups or cohorts become more specific, the number of units contributing to that specific cohort will decrease. Table 1 depicts an example of the various benchmarking cohorts that can be derived for one single unit.

Regardless of the number of benchmarks available to each area, the most important element is to identify the benchmarking cohort that is most reflective of the practice setting in order to establish the most accurate comparison group. Accurate benchmarks provide nurses with greater insight into the care they are providing by allowing them the ability to evaluate whether or not their nurse-sensitive outcomes are better, worse, or the same as others working in similar practice settings. Without evaluation of data and benchmarking, nurses would not be able to understand or fully appreciate the tremendous impact nursing practice has on patient outcomes. The continuous monitoring of data and benchmarking allows nurses to identify areas for improvement, collaborate with units/organizations that are superior performers, and implement best practices that will improve patient outcomes.

Table 1 Example of benchmarking cohorts

City Hospital is an academic medical center located in central Chicago, Illinois, in the United States. The hospital has 612 beds and has recently received Magnet Designation from the American Nurses Credentialing Center. The adult Medical Intensive Care Unit at City Hospital is looking to provide staff with comparison data related to catheter-associated urinary tract infections to evaluate how their current infection rates compare to those in the region, the state, nationally, and among similar types of hospitals. The Medical Intensive Care Unit would have access to the following benchmarking cohorts (where "*n*" = the number of units contributing to dataset):

- Adult Critical Care, All Contributing Hospitals ($n = 1009$)
- Adult Critical Care, Academic Medical Center ($n = 821$)
- Adult Critical Care, Magnet Hospitals ($n = 723$)
- Adult Critical Care, Bed Size Greater than 500 Beds ($n = 612$)
- Adult Critical Care, Urban Hospitals ($n = 425$)
- Adult Critical Care, Midwest Region ($n = 307$)
- Adult Critical Care, Illinois ($n = 45$)

The initial step in establishing a benchmark is to submit performance indicator data to a repository or database, either internally or externally. To ensure accurate comparable data and benchmarks are produced, it is essential that data abstraction and submission are conducted uniformly for each indicator. Data abstraction can be conducted through a centralized department, such as the organization's quality department, or decentralized through unit-level reporting. Centralized data abstraction often occurs when various units are reporting on the same performance indicator, such as patient falls or hospital-acquired infections, or when data are submitted to multiple benchmarking vendors or external sources, such as the Centers for Medicare and Medicaid Services (CMS) and the National Database of Nursing Quality Indicators (NDNQI). Decentralized data abstraction occurs when data are specific to only a handful of units, such as lactation consultant hours for a postpartum unit, or when unit-level audits are required to produce indicator data, such as pressure injury prevalence by unit type. Once data are abstracted and submitted to the repository or database from various organizations (or internal units of an organization when internal benchmarks are produced), data are analyzed and descriptive statistics of the comprehensive dataset are published. These descriptive statistics of the dataset become the "benchmarks."

Published benchmarking datasets most commonly include measures of central tendency, such as mean and median scores of the dataset, as well as various percentile rankings. Measures of central tendency are used to provide one value that describes a larger dataset, while percentile rankings allow organizations to determine the percent of other hospitals (or internal units) that scored better or worse than the hospital in question. Table 2 defines the key statistical measures in benchmarking. Using the example above from City Hospital, the Medical Intensive Care Unit reported an infection rate in the 90th percentile for the adult critical care, academic medical center cohort, which is supplied through an external benchmarking vendor. This signifies that the Medical Intensive Care Unit scored better than 90% of all the units within the comparison cohort of adult critical care units within an academic medical center that contributed patient infection data during the defined reporting period.

Both internal and external benchmarking systems will have a predetermined reporting period that defines the timeframe during which data are collected, typically monthly, quarterly, or annually. The benchmarks produced are representative of the performance of the contributing units

Table 2 Key statistical measures in benchmarking

	Definition	Example
Mean (average)	The sum of all the values in the dataset divided by the number of values in the dataset. (Measure of central tendency)	Dataset: 21, 28, 25, 32, 24 $\overline{x} = \frac{(21+28+25+32+24)}{5}$ Mean = 26
Median (50th percentile)	The middle value for a dataset that has been arranged in order of smallest to largest. (Measure of central tendency)	Dataset: 21, 28, 25, 32, 24 Order of magnitude = 21, 24, **25**, 28, 32 Median = 25
Percentile ranking	The percentage of values in the distribution of the dataset that are equal to or less than the value in question	Dataset: 21, 28, 25, 32, 24 32 is at the 80th percentile (4 of the 5, or 80%, of the values are equal to or less than 32)
Standard deviation	The measure of variation between values within a dataset	The lower the standard deviation, the closer the data points are to the mean, while the higher the standard deviation, the wider the range of values within a dataset

during that specific reporting period; therefore, it is important to remember that benchmarks will often fluctuate based on the data. Fig. 1 demonstrates how both the unit-level data (gray bars) and benchmarks (gray line) vary slightly each month (reporting period).

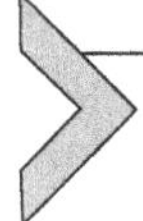

Common benchmarks, measurement, and reporting

Centers for Medicaid and Medicare Services

CMS is the largest payer for healthcare services in the United States and has a profound impact on all healthcare organizations across the country. In the early 2000s, legislation was passed within the United States that encouraged healthcare organizations to submit data to CMS for benchmarking, with the

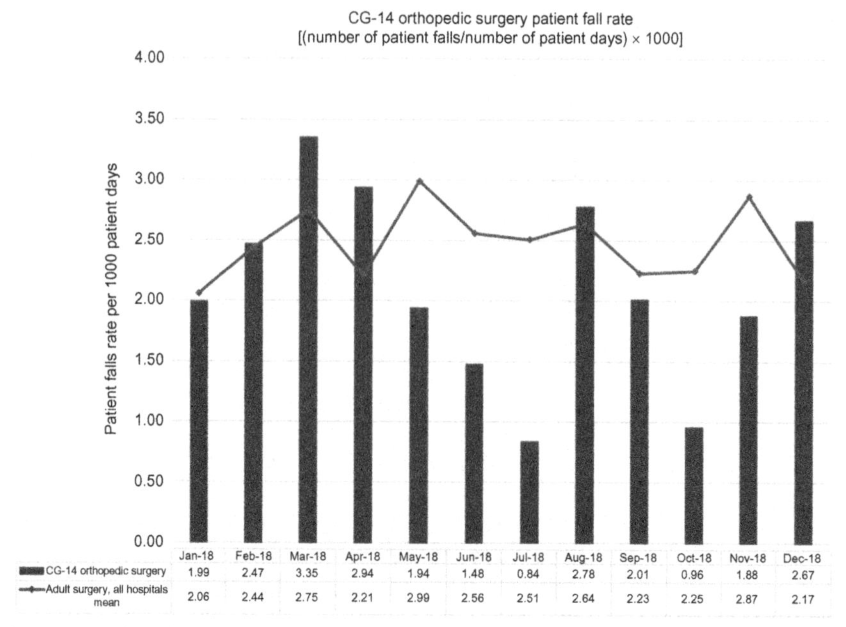

Fig. 1 Benchmark variability.

purpose of improving the quality of care and tying it to the amount of reimbursement these organizations received from CMS.[5] This early model has now evolved to one in which CMS can penalize healthcare organizations for poor performance on these identified benchmarks in efforts to maintain quality care delivery on a national scale. While this model is not necessarily established on an international level, there are significant benefits in having benchmarks set by a federal government for a country. The first benefit is a shared definition on how to measure and set a particular benchmark. When setting benchmarks, it is imperative to have a universal operational definition of how a measure is defined, measured, and interpreted. A second benefit of national, mandated benchmarks is that the federal government then has access to a large volume of standardized data across regions and populations. As the country's largest payer for healthcare in the United States, the CMS possesses large volumes of billing and coded data that are accessible for analysis. This collective data pool is an excellent source of information that allows for standardized hospital comparisons, informed decision making on future legislation, priorities, policies and costs, and remains a method to ensure quality of care across organizations.

CMS benchmarking is variable and changes occur on a regular basis. National benchmarking programs are frequently tied to legislative processes. In the United States, programs such as the Medicare Prescription Drug, Improvement and Modernization Act set many of the benchmarking definitions and targets.[6] Because these programs have such variability, healthcare organizations need to pay attention to legislative changes to stay on top of shifting benchmarking and targets. There are many benchmarking programs within CMS and other government agencies, but this section will focus on three of the primary programs, including CMS Core Measures, Hospital Compare, and Medicare Shared Savings Program (MSSP).

CMS Core Measures

The CMS Core Measures were designed to help hospitals, commercial payers, and patients establish a shared set of quality measures with uniform definitions and requirements. These measures were developed to be meaningful, easy to collect, and at a minimal cost.[7] Core measures are generally updated on an annual basis. CMS develops work groups to review and modify the core measure set. The National Quality Forum (NQF) is a nonpartisan, nonprofit organization that was created by leaders in the public and private sectors in the United States to generate evidence-based, consensus recommendations for measures to evaluate performance and improve healthcare. Recommendations for performance measures from NQF undergo a rigorous process for inclusion, and are key considerations when determining national benchmarks for quality by federal regulatory groups and accreditation bodies. Other groups such as the Healthcare Effectiveness Data and Information Set (HEDIS) and the National Committee for Quality Assurance (NCQA) also have an impact on the development of core measure sets. The most recent version of the core measures has eight different groups of measures: Primary Care, Cardiovascular, Gastroenterology, Human Immunodeficiency Virus (HIV)/Hepatitis C, Medical Oncology, OB/GYN, Orthopedic, and Pediatric. Though each group will have specific measures, many of the measures are shared among groups. Table 3 displays a listing of example measures for primary care and cardiovascular areas.[7]

Each of these items contains specific benchmarked measures that are compared at the level of the individual provider, or at the level of the facility. Data submissions vary by measures and program but generally occur quarterly or annually. Many of the benchmarks in the core measure set are used in different CMS programs, including Hospital Compare and Shared Savings Programs.

Table 3 Example quality measures for primary care and cardiovascular

Examples of primary care measures	Measure description
Controlling high blood pressure	Percentage of patients with a diagnosis of hypertension whose blood pressure is adequately controlled
Hemoglobin A1c (HbA1c) poor control (>9.0%)	Patients with diabetes whose most recent HbA1c level during the measurement period was greater than 9.0%
Cervical cancer screening	Patients who received the appropriate cervical cancer screening for their age group during the measurement period
Body mass index (BMI) screening and follow-up	Documentation of BMI and identified plan for BMI outside of normal parameters
Depression remission at 12 months	Patients with depression who are in remission at 12 months after new or existing diagnosis
Examples of cardiovascular measures	
Controlling high blood pressure	Percentage of patients with a diagnosis of hypertension whose blood pressure is adequately controlled
Coronary artery bypass graft (CABG) with beta blockade	Percentage of patient undergoing CABG discharged on beta blockers
Chronic anticoagulation therapy	Patients with atrial fibrillation with were prescribed an anticoagulant drug
Chronic stable coronary artery disease: Antiplatelet therapy	Patients with coronary artery disease prescribed aspirin or clopidogrel
Primary percutaneous coronary intervention (PCI) received within 90 minutes of hospital arrival.	Patients with acute myocardial infarction who receive percutaneous coronary intervention within 90 minutes of arrival

Hospital Compare

Hospital Compare is a benchmarking program intended to provide healthcare consumers a simple way of differentiating quality in healthcare organizations. It is a summary of 57 different quality measures in 7 different categories.[7] Table 4 displays an example of categories present at the time of publication.

Each of the categories focuses on a critical aspect of a patient's healthcare experience and outcome. These metrics and benchmarks are based on commonly occurring conditions and are not reflective of performance on complex or highly specialized care. Three categories are focused on care delivery

Table 4 Hospital compare benchmarking categories

Hospital compare categories	Example measure
Mortality: 7 measures	Death rates for the following patient groups: Heart attack patients, coronary artery bypass surgery patients, chronic obstructive pulmonary disease patients, heart failure patients, pneumonia patients, stroke patients Also includes a measure for the number of deaths among patients with serious treatable complications after surgery
Safety of care: 8 measures	Catheter-associated urinary tract infections, central line associated blood stream infections, surgical site infections from colon surgery or abdominal hysterectomy, methicillin-resistant staphylococcus aureus (MRSA) bloodstream infections, clostridium difficile infections, complication rates for hip/knee replacements and other serious complications
Readmissions: 9 measures	Rate of unplanned readmission after: Hip/knee surgery, heart attack, coronary artery bypass graft surgery, chronic obstructive pulmonary disease, heart failure, pneumonia, stroke Also includes overall rate of unplanned readmissions hospital wide and rate of unplanned hospital visit after outpatient colonoscopy
Patient experience: 11 measures	Patient reported that: nurses communicated well, doctors communicated well, received help as soon as they needed it, received explanations about medications, had clean rooms, quiet environment at night, received homegoing instructions, understood care when leaving the hospital Includes additional measure for overall hospital rating
Effectiveness of care: 10 measures	Percentage of patients who left the emergency department before being seen, administration of influenza vaccine to patients, administration of influenza vaccine to healthcare workers, number of patients who received aspirin for chest pain within 24 hours, percentage of patients with suspected stroke who received brain scan results within 45 minutes, patients received appropriate recommendation for follow up for screening, colonoscopy patient with polyps received follow up colonoscopy, percent of deliveries scheduled

Continued

Table 4 Hospital compare benchmarking categories—cont'd

Hospital compare categories	Example measure
	too early among pregnant women, percent of patients receiving appropriate care for sepsis, percentage of patients who developed preventable blood clot, percentage of patients receiving radiation therapy for bone cancer
Timeliness of care: 7 measures	Average time patients spent in the emergency department before they were admitted to the hospital as an inpatient, average time for blood clot medication administered to heart attack patients, average time heart attack patients were transferred to necessary care, average time for chest pain patients to receive ECG, time spent in emergency department before being seen, average time until patients with broken bones were administered pain medication
Efficient use of medical imaging: 5 measures	Outpatient CT scans of the abdomen that were "combination" (double) scans, outpatient CT scans of the chest that were combination scans, outpatients with low back pain who received MRI without other treatments first ,outpatients who received cardiac stress imaging before low risk outpatient surgery, outpatients with brain CT scans who received sinus CT scans at the same time

and outcomes, and include: Mortality, Safety of Care, and Readmissions. Another category centers on patient perception of their healthcare experience. This information is obtained through the Hospital Consumer Assessment of Healthcare Provider and Systems (HCAHPS) Survey. This survey assesses how well healthcare staff communicate with patients, the impression of a hospital environment, and how well a healthcare provider responded to patient needs.[7] These first four groups of measures are weighted heavily in overall calculations. Collectively they represent 88% of the total Hospital Compare rating. The final three groups of measures include Effectiveness of Care, Timeliness of Care, and Efficient use of Medical Imaging. These groups of measures are generally focused on efficiency and represent 12% of the total Hospital Compare rating. Healthcare organizations frequently allocate significant quality improvement resources on these four groups of measures.

Medicare Shared Savings Program

The Medicare Shared Savings Program (MSSP) is another benchmarking program within the United States that allows healthcare organizations to develop care programs known as Accountable Care Organizations (ACO).[6] ACOs are shared risk programs that financially reward organizations for decreasing costs, while also assessing penalties to organizations that are not meeting assigned benchmarks. These programs are optional for organizations to participate in, but are becoming more common as participation in them continues to demonstrate decreases in healthcare costs and retain a focus on care value and quality. CMS sets the MSSP program benchmarks every 2 years. In 2018–19, the program consisted of measures focused on four different quality areas: Patient/caregiver experience, care coordination and patient safety, preventive health, and at-risk populations. Table 5 displays the measures under each quality area.

For each of the MSSP measures, CMS assigns specific benchmarks. Based on an organization's performance on each measure, points are then awarded. The total number of points at the end of a calendar year determines if an organization gets additional healthcare dollars, or if penalty fees are assessed.

Table 5 Medicare shared savings program quality measures

Measure domain	Example measure
Patient/caregiver experience: 10 measures	Consumer Assessment of Healthcare Providers and Systems (CAHPS): How well your providers communicate, receiving timely care, provider rating, access to specialists, health promotion and education, shared decision making, health and functional status, stewardship of patient resources, courteous and helpful office staff, overall care coordination
Care coordination/patient safety: 4 measures	All-cause unplanned admissions for patients with multiple chronic conditions and overall all condition readmission. Ambulatory sensitive condition prevention. Screening for future fall risk
Preventative health: 6 measures	Breast cancer screening, colon cancer screening, depression screening, statin therapy for cardiovascular disease, influenza immunization, tobacco screening and cessation intervention
At-risk populations: 3 measures	Depression remission at 12 months, hemoglobin A1c control for diabetic patients, hypertension control

Accreditation organizations

Benchmarking is not done solely by governmental agencies within a specific country. Accreditation bodies like The Joint Commission also provide benchmarking opportunities. Specifically for The Joint Commission benchmarks, measures can focus on system-wide factors, or disease-specific items. For consistency, The Joint Commission utilizes measures identical to the CMS Core Measures. Data submission on these specific measures demonstrates integration of performance management techniques into the accreditation process for hospitals and health systems. These data are required to be submitted quarterly and for hospital accreditation.

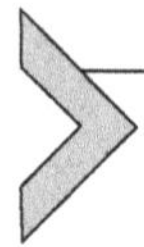

Nursing specific benchmarks, measurement, and reporting

In addition to national reporting of benchmarks, organizations may also voluntarily choose to submit data for further benchmarking. Voluntary reporting acts as an added resource to aid in the identification of best practices and continuous quality improvement. Nurses often use voluntary reporting to benchmark nurse-sensitive indicators. Nurse-sensitive indicators are outcome, structure, or process measures that are reflective of the nursing care provided and directly impact quality of care.[8] These nurse-sensitive indicators help guide decision making, the identification of improvement needs, and the allocation or reallocation of resources. The added element of benchmarking allows nurse leaders to advocate for additional resources to support nursing practice. For example, this can be demonstrated through the application of research that supports the correlation between hospital-acquired condition or other adverse event occurrences and poor staffing ratios.[9] Nurse leaders that identify underperformance of national benchmarks in these nurse-sensitive indicators can use data and benchmarking to advocate for increased staffing to decrease nursing workload and improve quality of care.

Although there are multiple benchmarking vendors that identify specific nurse-sensitive indicators, the National Database of Nursing Quality Indicators (NDNQI) is one of the most widely used databases for nurse-sensitive indicators and includes outcome, structure, or process measures. Outcome measures evaluate quality and quantity of nursing care; structure measures evaluate the supply and skill/educational demographics of the nursing staff; and, process measures evaluate elements of nursing care, such as nursing interventions, practice, and satisfaction.[8] NDNQI was founded by the American Nurses Association (ANA) as part of the ANA's Safety and Quality Initiative and was built upon the success of multiple pilot studies conducted by ANA throughout various states.[8] The NDNQI database was acquired by Press Ganey Associates, Inc., in 2014 which has allowed organizations to enhance data analytic and reporting capabilities specific to nurse-sensitive indicators. Table 6 identifies the various clinical measures benchmarked by NDNQI. The vendor also provides benchmarking capabilities for nurse satisfaction.

The Collaborative Alliance for Nursing Outcomes (CALNOC) is another leader in the healthcare industry for nurse-sensitive indicator

Table 6 NDNQI clinical measures[8]

Outcome measures	Assaults by psychiatric patients
	Assaults on nursing personnel
	Catheter-associated urinary tract infections (CAUTI) rate
	Central line catheter associated blood stream infections (CLABSI) rate
	Hospital readmissions rate
	Multidrug-resistant organisms (MDRO)
	Patient falls rate and injury falls rate[a]
	Pediatric peripheral intravenous infiltrations
	Pressure injuries (hospital-acquired and unit-acquired rates)[a]
	Ventilator-associated events (VAE) rate
	Ventilator-associated pneumonia (VAP) rate
Structure measures	Nurse turnover
	Patient volume and flow (admissions, discharges, and transfers and throughput)
	RN education
	RN specialty certification
	Staffing and skill mix (hours per patient day, nursing care hours, nursing care minutes)
Process measures	Care coordination
	Device utilization (central lines, catheters, ventilators)
	Pain-impairing function
	Patient falls[a]
	Pressure injuries[a]
	Pediatric pain assessment/intervention/reassessment (AIR) cycle restraints

[a] Patient falls and pressure injury reporting have elements of both outcome measures (the actual rate at which patients falls or prevalence of pressure injuries), as well as process measures, such as compliance with interventions to prevent patient falls or pressure injuries.

benchmarking. CALNOC originated as one of the ANA pilot study sites that influenced the design and creation of NDNQI and grew into an independent, nonprofit corporation dedicated to providing nurses with access to benchmarking data for nurse-sensitive indicators.[10] In 2019, Press Ganey announced the acquisition of the CALNOC database to further expand benchmarking capabilities, as well as nurse-sensitive indicators in the ambulatory setting. Table 7 identifies the wide variety of clinical measures benchmarked through CALNOC.

As healthcare delivery models have shifted within specific countries, and on an international scale, the role of nurses in ensuring care quality continues to adapt as well. Healthcare organizations are shifting priority areas from highly complex specialty care to a focus on preventative care and population health initiatives that aim to keep individuals out of the hospital, and decrease the complexity of disease progression and management. The nursing role within ambulatory and clinic settings now may include care coordination, patient education, and promotion of preventative screenings that will help bridge the gap between coverage and access and provide highly coordinated, quality care.[11] With these shifts emerged the need to assess

Table 7 CALNOC clinical measures[10]

Outcome measures	Hospital-acquired pressure ulcer by stage[a]
	Patient fall rate and patient injury fall rates[a]
	Restraint prevalence rate
	Central line-associated blood stream infections in PICC lines
	Medication administration accuracy nurse safe practices finding and error rates[a]
Structure measures	Hours of nursing care per patient days
	Skill mix
	Contracted employee hours
	Staffing ratios
	Voluntary turnover
	RN characteristics—Education, experience, years of service
	Unit rate of admissions, discharges, and transfers
Process measures	Patient falls[a]
	Hospital-acquired pressure ulcers[a]
	Medication administration accuracy safe practices[a]
	PICC line insertion practices

[a] Patient falls, pressure injury, and medication administration reporting have elements of both outcome measures (the actual rate at which patients fall, acquire pressure injuries, or encounter medication errors), as well as process measures, such as compliance with interventions to prevent patient falls, pressure injuries, and medication administration errors.

nursing impact on care coordination and identify specific performance metrics that accurately depict the role of the nurse in the ambulatory practice settings.

In response to these shifts, the American Academy of Ambulatory Care Nursing (AAACN) and CALNOC have been working collaboratively since 2013 to develop meaningful nurse-sensitive indicators and benchmarking measures for ambulatory care practice settings. In 2016, the collaborative released five nurse-sensitive indicators for Ambulatory Surgery Centers in the United States, and in 2017, additional indicators were released relative to other ambulatory practice settings such as primary care, specialty care, urgent care, and birthing centers.[12] Table 8 lists the ambulatory nurse-sensitive indicators as suggested by AAACN and CALNOC. AAACN and CALNOC continue to work to develop more complex performance measures such as Care Coordination, Complex and Chronic population group metrics, and Tele and Virtual Health metrics, as well as advancing the care planning metrics. Additionally, NDNQI has released two care coordination indicators (medication reconciliation and pending diagnostic test results), and created additional benchmarking capabilities within the ambulatory practice setting to better capture nursing contributions in this area. With Press Ganey's combined management of the NDNQI and CALNOC databases, nurses are likely to gain access to an even broader range of nurse-sensitive indicators and benchmarking, particularly within the ambulatory setting.

Table 8 Ambulatory nurse-sensitive indicators

Ambulatory setting type	Nurse-sensitive indicator
Surgery and procedural	Staffing, skill mix, and patient care hours
Primary care	Visit volume
Specialty care	Adverse outcomes of care[a]
Urgent care	Patient burns[a]
Birthing center	Patient falls and injury falls
	Hospital transfer/admission
	No show/cancellation[b]
	Pain assessment and follow up[c]
	Hypertension assessment and follow up
	Community falls risk and follow up
	Body mass index assessment and follow up
	Depression assessment and follow up

[a] Applicable to surgery and procedural.
[b] Not applicable to urgent care.
[c] Not applicable to birthing center.

Conclusion

Internal and external benchmarking provides organizations the opportunity to evaluate quality and value of healthcare both within and across organizations and care delivery settings. Nurses continue to play an integral role in care delivery and must be knowledgeable of benchmarks within their organizations, regions, and counties. Many of these benchmarks are directly impacted by nursing interventions and type of care provided. Ongoing surveillance of these measures is important for nurses in order to assess effectiveness of changes in real time, and integrate efforts to optimize outcomes for cost-effective care. Comparison of measures across organizations allows for ongoing monitoring of factors affecting care quality on a regional, national, and international scale and how factors impact health of a population. Nurses remain a central component to care delivery, impacting benchmarking data, and thus must remain knowledgeable of current benchmarks to advance quality care, ensure safety, and optimize outcomes for patients.

References

1. Vlăsceanu L, Grünberg L, Pârlea D. *Quality Assurance and Accreditation: A Glossary of Basic Terms and Definitions*. Bucharest: UNESCO-CEPES; 2007. Retrieved from: http://unesdoc.unesco.org/images/0013/001346/134621e.pdf.
2. Ettorchi-Tardy A, Levif M, Michel P. Benchmarking: a method for continuous quality improvement in health. *Healthc Policy*. 2012;7(4):e101–e119.
3. National Quality Forum. *National Voluntary Consensus Standards for Nursing-Sensitive Care: An Initial Performance Measure Set*. Washington, DC: National Quality Forum; 2004.
4. National Research Council. *Measuring Performance and Benchmarking Project Management at the Department of Energy*. Washington, DC: The National Academies Press; 2005. https://doi.org/10.17226/11344.
5. Brenner ZR, Salathiel M. The nurse's role in CMS quality indicators. *Medsurg Nurs*. 2009;18(4):242–246.
6. Centers for Medicare & Medicaid Services. *Quality Measures*. Retrieved from: https://www.cms.gov/Medicare/Quality-Initiatives-Patient-Assessment-Instruments/QualityMeasures/index.html; 2018.
7. Centers for Medicare and Medicaid Services. *Core Measure*. Retrieved from: https://www.cms.gov/Medicare/Quality; 2018.
8. Press Ganey Associates, Inc. *General Overview: Introduction to Guidelines for Data Collection and Submission on Indicators*. Retrieved from: https://members.nursingquality.org/NDNQIPortal/NDNQI/Documents.aspx; 2018.
9. Cho E, Chin D, Kim S, Hong O. The relationships of nurse staffing level and work environment with patient adverse events. *J Nurs Scholarsh*. 2016;1:74–82.

10. CALNOC. *Overview*. Retrieved from: https://www.calnoc.org/page/15; 2017.
11. Institute of Medicine. *The Future of Nursing: Leading Change, Advancing Health*. Washington, DC: National Academies Press; 2010.
12. Start R, Matlock AM, Mastal P. *Ambulatory care nurse-sensitive indicator industry report: Meaningful measurement of nursing in the ambulatory patient care environment*. Pitman, NJ: American Academy of Ambulatory Care Nursing; 2016.

CHAPTER THREE

Research data in hospitals and health systems

Joyce Karl, Lorraine Mion
College of Nursing, The Ohio State University, Columbus, OH, United States

Contents

Importance of research data in health systems

Research in health systems provides data for administrators and clinicians to identify effective practices for specific patient populations in order to streamline resources and positively impact patient and system outcomes. Data from research can challenge beliefs or misconceptions that have often been held as "truth" or "common sense" among clinicians. For example, nurses use bed alarms and nonskid socks to prevent falls; yet the evidence does not support widespread adoption of these practices.[1, 2] Moreover, these practices consume staff resources, and can lead to increased falls directly or indirectly by creating a false sense of security. Similarly, nurses would historically report they had to use physical restraints to prevent falls or prevent patients from removing medical devices. However, not only did research demonstrate that patients still fall or remove devices with physical restraints

Data for Nurses
https://doi.org/10.1016/B978-0-12-816543-0.00003-0

in place, but that the involuntary immobilization due to restraints can lead to adverse consequences, including death. As a result, these data were the impetus for changes in accreditation and federal regulations regarding restraint usage in health systems.[3, 4] While research will never replace sound professional judgment, clinical experience, and other sources of data, it does provide a mechanism to question practices and impact on patient and organizational outcomes.

Differences between research and quality data

Before discussing research data and interpretation of research data, it is important to summarize the differences in quality improvement (QI) and research methods. QI and research designs can overlap, creating a degree of uncertainty of whether a project is QI or a project is research. QI and research procedures differ in focus, goals, and designs.[5]

Typically, QI focuses on the internal processes and practices within a health system in order to continually improve on the outcomes and efficiency of the organization. The types of interventions or strategies implemented in QI projects are ones that have been well tested and shown in research literature to be effective. Thus, risks to patients should be minimal or absent since the practices are considered best practices and are intended for all patients. Protections are in place to maintain the anonymity and confidentiality of any QI data that are collected. Data are collected without patient or practitioner identifiers and are reported in aggregate. The knowledge generated is specific to the organization and thus tends to be of limited generalizability to other settings.

In contrast, the focus of research is to generate new knowledge with the ultimate goal to guide practice and/or policy. There are numerous designs in research. Qualitative designs, such as focus groups or ethnographic studies, seek to understand the patient's or nurse's perspectives. The type of data generated provides themes or patterns, but no numeric values. Quantitative designs can be observational (e.g., descriptive, correlational, cohort, case-control), in which the main focus is to understand the scope of occurrences, patterns of occurrence, relationships among variables, or risk factors and outcomes. The investigator does not manipulate any of the variables but collects data as they occur naturally. Experimental designs can range in degree of rigor, from one group pre-post quasiexperiments to the gold standard of randomized-controlled trials. The data collected in these types of studies provide not only descriptive statistics on frequencies, central tendency measures (e.g., mean, median), correlations, and predictors, but also examine the effect of an intervention, which the investigator manipulates. The major difference between QI data and research data is that research data are based on *individual* values, rather than on aggregate values.

Interpretation of research data

It is beyond the scope of this chapter to provide a complete text on statistical data, reporting, and interpretation. A brief review is provided, and readers are encouraged to seek expert resources when unsure of how to interpret research data. Clinicians and administrators must understand research data and not rely solely on authors' conclusions before enacting changes in the organization and delivery of care. Indeed, many articles continue to have errors in reporting statistics, at times serious enough to call into question the authors' conclusions.[6, 7] Table 1 provides a quick reference of the types of research data.

Table 1 Descriptions and rationale for commonly reported research data: Descriptive statistics[7]

Term	Description
Categorical variables descriptors	
Frequency	A count of occurrences. Example: 20 men out of 100 participants.
Ratio	Juxtaposition of two numbers indicating how many times the first number contains the second. For example, 1:5 ratio of men to women would be interpreted as 1 man for every 5 women. There is no arithmetic relationship between a numerator and a denominator.
Proportion	Numerator/denominator. For example, 20 men/100 participants = 0.2
Percentage	Numerator/denominator × 100%. Allows comparison of groups of different sizes. For example, 20/100 × 100% is 20% men.
Risk	Probability of an event occurring.
Relative Risk	Probability or relationship of two events. Used in prospective cohort studies. For example, risk of MI for smokers compared to MI for nonsmokers.
Odds	Likelihood of an event occurring as compared to not occurring.
Odds Ratio	Ratio of two odds. Used in cohort studies and case-control studies.

Continued

Table 1 Descriptions and rationale for commonly reported research data: Descriptive statistics[7]—cont'd

Term	Description
Continuous variables descriptors	
Mean	Sum of values divided by total number of participants/units.
Median	The value at the midpoint of all observed values.
Mode	Value that appears most often.
Standard deviation	The square root of the average of the squared differences from the mean. For (approximately) "normally" distributed, i.e., "bell-shaped" curve distributions. Meaningless if not normally distributed.
Range	Maximum value–minimum value.
Interquartile range	75th percentile value–25th percentile value. Minimizes influence of outliers.

When interpreting research data, consider the level of the variable. Categorical variables consist of nominal and ordinal variables. There is no ranking of the values within a nominal variable. Examples include gender (male or female) or political party affiliation (Republican, Democrat, Libertarian, Independent, Other). The next level is ordinal. There is a ranking of the values from lowest to highest, but the preciseness of amount of difference from one value to the next is absent. Examples include level of function (independent, requires minimal help, requires moderate help, dependent) or Likert scales, such as level of satisfaction (very dissatisfied, somewhat dissatisfied, neither satisfied nor dissatisfied, somewhat satisfied, very satisfied). Continuous or interval variables are ones in which the intervals between the values are the same. For example, inches on a ruler meets this definition; the distance between 2 and 3 in. is the same as the distance between 7 and 8 inches.

Descriptive data[7]

Descriptive data are summaries. Frequency, proportion, percentage, rates and/or ratio can be used for nominal and ordinal variables. Central tendency measures are provided for continuous variables and ordinal variables in some instances, e.g., seven-point Likert scales. These reports summarize the distribution of the values along a continuum. Common central tendency measures are mean, median, and mode. Measures of dispersion provide the spread of the distribution and include standard deviation, range, or interquartile range.

Inferential statistics[6]

Inferential statistics allows one to draw conclusions that extend beyond the sample studied. Inferential statistics are used to test a hypothesis, derive estimates, gauge the strength of associations, or determine level of risk or prediction. As with descriptive statistics, the type of variable as well as the type of group(s) (independent or paired) will guide the choice of statistical test. When interpreting the test result, one will typically see *P*-values, confidence intervals, or odds ratios and risk estimates. There are a few points to consider with each of these.

Statistical significance does not necessarily mean clinical significance

A *P*-value, such as .05 or .01, can indicate a statistically significant finding, which is based on probability. That is, what is the likelihood that the finding is due to chance rather than a true observation? A $P\text{-value} = .05$ means that there is a 1 in 20 chance that the observed finding is due to chance. *P*-values are influenced by the sample size; very large sample sizes can result in statistically significant results for small differences or correlations (i.e., effect size). Clinically significant findings, however, are based on meaningful effect sizes. It is important to examine the size of the difference or correlation since large sample sizes can produce a statistically significant finding for small differences or correlations. For example, a researcher reports that among 2500 adults with anxiety, using lavender aromatherapy resulted in a statistically significant decrease in systolic blood pressure: from a mean of 146 to a mean of 144. As a clinician, you must determine whether this mean difference is clinically significant and warrants the resources to implement a change in practice.

Confidence intervals

There is always some degree of error in statistics since they estimate the findings from a sample to a population of interest. To allow one to interpret the reported research data, confidence intervals of 90%, 95%, or 99% are often provided. Confidence intervals (CI) are a range of values that purport to encompass the "true" value and are used descriptively or inferentially. The 95% confidence interval is related, but not the same, as statistical significance at the .05 level. For example, we conducted a study on a cohort of ventilated medical intensive care unit (MICU) adults examining the prevalence of severe agitation.[8] Severe agitation occurred in 16.1% (95% CI: 10.9%–23.0%) of the adult ventilated patients. One would read this as:

among this sample, 16.1% experienced severe agitation and there is a 95% chance that this particular confidence interval (10.0%–23.0%) contains the true population frequency of severe agitation among ventilated MICU patients. The second piece of information provided by the CI is precision of the estimate, the narrower the width, the more precise the estimate. Third, when examining differences between groups or in the same group over time, the CI is statistically significant if the range does not include zero. For example, the mean difference in fall rates was 0.8 with a 95% CI of −0.20 to 1.4/1000 patient-days (not significant). This would be interpreted as: for this study sample, fall rates decreased by a mean of 0.8 but the CI indicates that rates could have increased by 0.20 or decreased by 1.4. Fourth, significant CI intervals for odds ratios or risk ratios do not contain 1; ratios that are greater than 1 indicate increased risk for one group over another; a ratio less than 1 means decreased risk. For example, we found in our sample that a PaO_2/FIO_2 <200 mg had a hazard ratio of 1.61 (95% CI: 1.02–2.54) for severe agitation. Thus, patients with those values were 1.61 times more likely to develop severe agitation. Since both values in the range are >1, then this would be considered a significant increase. If both values in the range were <1, then it would be considered as predictive of *not* getting agitation.

Models of evidence-based practice

The next step in use of research data is determining whether and how to integrate these data into nursing practice. Initially, "research utilization" described the application of research data into clinical practice. The term "evidence-based practice" includes not only the use of best available research evidence but also considers internal evidence (i.e., the patient's status and conditions, QI or outcomes management data), patient preferences, healthcare resources, and clinical expertise.[9, 10] Evidence-based practice (EBP) implementation requires knowledge and skills.[11] One must have the ability to formulate and ask the clinical question to guide the appropriate literature search. One must then comprehend and interpret research findings and determine whether the findings are applicable to the population and clinical practice of interest. In determining the strength of evidence, one must understand similarities and differences among research, EBP, and QI. Last, one must have the ability to accurately and appropriately collect, interpret, use and disseminate data within the organization to gauge the success of the EBP project. There are a number of models or frameworks to guide EBP clinical decision making and processes to change practice.[9] We present several of the more commonly used frameworks to guide nursing practice.

The Stetler Model of evidence-based practice[9, 12]

The Stetler Model was the first nursing model for research utilization, published in 1976. Subsequent revisions in 1994, 2001, and 2009 built on the need to use research knowledge in clinical practice, integrate EBP concepts, and refine essential concepts including the addition of external and internal sources of evidence. External evidence includes original research findings as well as expert/consensus opinion of well-known experts or reliable program evaluations in peer-reviewed literature when findings are deficient. Internal evidence comes from local, methodically acquired data consensus/opinion of local groups or experiences of individual professionals/experts. The Stetler Model is considered a practitioner-oriented model because it focuses on critical thinking and use of research and evidence at the individual professional level.[9]

EBP process models

Several conceptual models have been developed over the years to guide EBP processes and implementation (Table 2). These models have various

Table 2 Common evidence-based practice models

Model	Type	Key concepts/features
Stetler Model for research utilization[12]	EBP, RU, and knowledge transformation (process)	Five phases to use of research data: (1) preparation, (2) validation, (3) comparative evaluation/decision making, (4) translation/application, and (5) evaluation
The Iowa Model of evidence-based practice[13]	(process)	Starts with encouraging clinicians to identify triggers/clinical issues. Uses a multiphase change process with feedback loops to guide clinical decisions and organizational practices that affect patient care outcomes

Continued

Table 2 Common evidence-based practice models—cont'd

Model	Type	Key concepts/features
Model for evidence-based practice change[14]	(process)	Guides practice change using six steps: (1) assess need for practice change, (2) locate best evidence, (3) critically analyze the evidence, (4) design practice change, (5) implement and evaluate change in practice, (6) integrate and maintain change in practice
The Advancing Research and Clinical practice through close Collaboration (ARCC) Model[15]	(system-wide and process)	Evidence-based system-wide implementation and sustainability model for EBP. Encompasses strategies for individual clinicians and organizational change. Uses EBP mentors to facilitate evidence-based care with clinicians and to create a culture that supports and sustains EBP. Model use has studied and supports improved clinician and patient outcomes
The Promoting Action on Research Implementation in Health Services (PARHIS) framework[16]	(system-wide)	Successful implementation requires three main elements: (1) evidence (e.g., rigorous studies, clinician's expertise/experience, patient's/caregiver's expertise), (2) context, and (3) facilitation. When each of these elements is at a high level, the probability of successful EBP implementation is enhanced.[6]

Modified from Dang D, Melynk BM, Fineout-Overholt E, Ciliska D, Cullen L, Cvach M. Models to guide implementation and sustainability of evidence-based practice. *Evidence-Based Practice in Nursing & Healthcare: A Guide to Best Practice*. 3rd ed. Philadelphia, PA: Wolters Kluwer Health; 2015:274–315.

strengths and can be valuable for a variety of purposes. Most of the models, including the Stetler Model, are process focused and include steps to implement evidence-based care.[15] General features include (a) identifying a clinical problem or issue that needs to be addressed, (b) finding and critically appraising the existing evidence for quality and fit, (c) identifying stakeholders, facilitators, and barriers, (d) implementation strategies for practice change, (e) evaluating the impact of change, and (f) identifying strategies to sustain and re-evaluate the change.

EBP system-wide models

System-wide EBP models provide a structured framework to guide implementation and sustainability of EBP throughout the healthcare organization, improve quality of care processes, and improve patient outcomes (Table 2). We present two system-wide models: the Advancing Research and Clinical practice through close Collaboration (ARRC ©) Model and the Promoting Action on Research Implementation in Health Services (PARIHS) framework.[15–17] Table 2 provides the reader with additional established EBP models that may be a fit for one's aims, purposes, or settings.[9]

ARRC model

As with a number of EBP models, the ARRC Model provides steps in the EBP process[18]:

Step 0: Cultivate a spirit of inquiry.

Step 1: Ask clinical questions in the PICOT format: Patient population (P), Intervention or area of interest (I), Comparative intervention or group (C), Outcome (O), and if applicable, Time (T).

Step 2: Search for the best evidence using the PICOT format.

Step 3: Critically appraise the evidence.

Step 4: Integrate the evidence with clinical expertise and patient preferences and values.

Step 5: Evaluate the outcomes of the practice decisions or changes based on evidence.

Step 6: Disseminate EBP results (to key stakeholders).

The ARRC Model goes beyond the individual clinician by incorporating an organizational assessment of the EBP culture to identify readiness, strengths, and potential barriers to EBP implementation within the organization. ARRC also features the development and use of EBP mentors in sufficient numbers to assist point-of-care clinicians in not only recognizing the value of EBP care, but also in enhancing their confidence in implementing EBP care.[15]

PARIHS framework[16, 19, 20]

The PARIHS framework accounts for the complex, multilevel, and dynamic interplay and interdependence of multiple factors that affect the successful implementation of EBP. As with the ARRC Model, PARIHS moves beyond the individual clinician level. Three major domains of evidence, context, and facilitation are incorporated that are on a continuum of high to low strength. Evidence includes not only research data, but data from a variety of sources, including the patient and organizational data. Context refers to the setting in which the proposed change is to be implemented and incorporates the culture, leadership, and evaluation. Facilitation refers to the process to enable the change to take place as well as the person (facilitator) who helps others in making the changes.

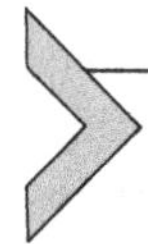

Relationships among research, evidence-based practice, and quality improvement data

Research, EBP, and QI data are interwoven and data from one approach often direct another approach. Gallagher-Ford provides a model that aligns not only research, EBP, and QI but also innovation (Fig. 1). The model displays that questions are the initiating force behind EBP. These questions can arise from organizational leadership or the individual clinician. Questions can also arise from the organization's performance metrics (QI). One must be cautious that QI processes monitor what we know to be best practices; do not mistake the QI process for the outcome itself. Results from the literature search determine the best approach. Insufficient evidence would lead to research to support the change and innovation. If there is sufficient evidence, then one implements the change using EBP processes and QI to monitor and sustain the EBP.

It is important to note that EBP overlaps with both research and QI, yet retains innate qualities that also distinguishes it as different (Table 3). EBP focuses on whether the best practices are in place, whereas research determines the best practice, and QI examines whether the best practices are done effectively. Unlike research that generates new knowledge, EBP takes the knowledge generated from research, and uses an implementation model or framework to implement the evidence into practice. Thus, similar to research, EBP may require institutional review board approval for data privacy and confidentiality protections, and if dissemination of implementation results in publication or presentation at professional meetings.

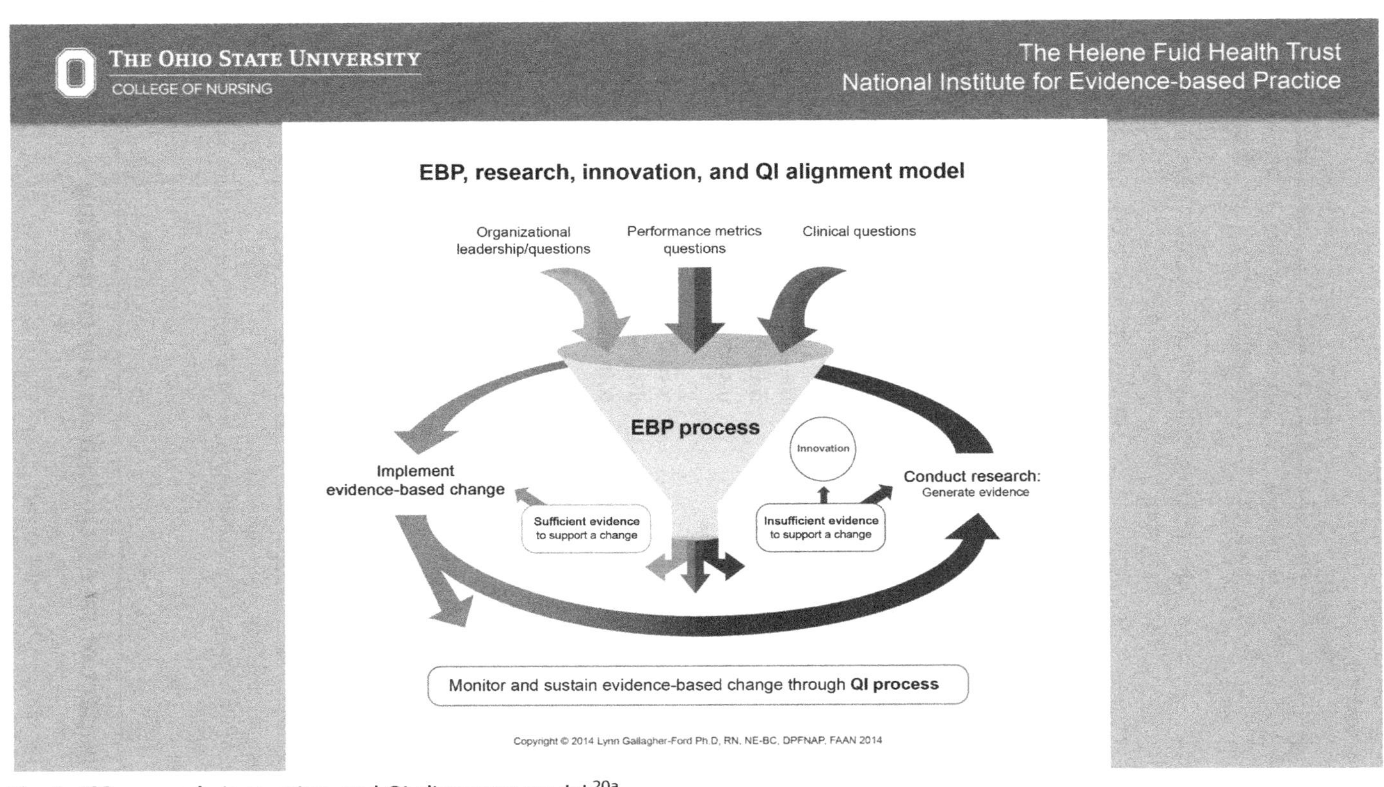

Fig. 1 EBP, research, innovation, and QI alignment model.[20a]

Table 3 Distinguishing research, evidence-based practice (EBP) or quality improvement (QI)

	Research	EBP	QI
Question	"What is the best thing to do?"	"Are we doing the best thing?"	"Are we doing the best thing effectively?"
Process All start with significance of problem	Search the literature to determine the gap Conceptual framework Hypothesis generating or hypothesis testing	Search and appraise the evidence Implement best evidence for practice change using an implementation framework Measure outcome of practice change	Examine internal evidence Ongoing implementation of best evidence for practice change Measure outcomes of practice change
	Generate new knowledge	**Implement evidence**	**Continue to improve care**
	Disseminate findings internally and externally	**Disseminate findings internally**	**Disseminate findings internally**
		External dissemination in some situations	**External dissemination in some situations**
Techniques	Research methods (e.g., randomized controlled trials, or observational designs)	PICO(T), ARCC, PARIHS, Rosswurm and Larrabee, Iowa, Hopkins models	PDSA, Six Sigma, TQM, Lean process, Dashboards, Scorecards
IRB or other formal review process	Yes (for protection of human subjects)	Yes (for protection of data and publication)	Yes (for publication)

Bold highlights differences in process and focus.
Based on the Organizational On-Site EBP Immersion Workshop (https://fuld.nursing.osu.edu/on-site-ebp-immersions).

Impact of research data on benchmarking

Research does not occur in isolation. Rather, research data are relevant for policy and practice, as well as directing future research. As discussed in earlier chapters, benchmarking allows comparison of both processes and outcomes over time within an organization, and the ability to compare to similar units or health systems. Benchmarking serves as a basis for quality improvement as well as self-regulation, and has become essential for reimbursement and accreditation and for avoidance of penalty fees.

Research data can provide the organization and clinicians with guidance on how to approach measurement (e.g., the right questions to ask), standardization of measurements, and establishing comparative metrics. Hospital falls is a primary example of this concept. For the conclusion of this chapter, two case examples are included to highlight the relationships between research, EBP, and QI in the clinical setting. The first example provides an example of using research data to evaluate frequency of falls within a health system. The second example illustrates the synchronous relationship between research, EBP, and quality within a health system using catheter associated urinary tract infections as a clinical exemplar.

Case example: Example of research data on falls frequency

Your unit, a medical stepdown unit with a high proportion of older adults and acute neurology conditions, has the highest rate of falls in the hospital at 7.3 fall events/1000 patient-days. Hospital administration has been pressuring you and your staff to reduce falls to the hospital benchmark of 2.3 fall events/1000 patient days. Despite your efforts in instituting evidence-based fall prevention practices, your unit's fall rates remain high. You examine the literature and find that studies have reported similarly high rates of falls in these particular units, among patient populations that are similar to yours. You also find that all fall events may not be the best benchmark to use with your patient population; rather, the rate of assisted falls versus unassisted falls would better demonstrate the use of best practices in (a) identifying those likely to fall and (b) staff assisting patients in ambulation and transfers. You apply this benchmark to your data, and find that 82% of all fall events on your unit are assisted fall events. Last, you determine that your fall injury rate is actually the lowest in the hospital at 0.8 fall injuries/1000 patient days. Therefore, while your overall fall rate may be high, most of those falls are actually assisted, meaning that staff are present, and that hardly any of those falls result in injury. Armed with these internal quality improvement data, evidence-based practice information, and external research data, you are able to demonstrate to hospital administrators that best practices on your unit are being done, and are resulting in quality outcomes.

Case example: Research-EBP-QI exemplar: Catheter associated urinary tract infection (CAUTI)

In the United States, catheter-associated urinary tract infection (CAUTI) is the most frequent hospital-acquired infection, comprising approximately 40% of all healthcare-associated infections.[21–23] CAUTI is a major cause of increased morbidity, mortality, and cost globally in hospitals and in long-term care.[21, 24, 25] A significant body of *research* has been conducted that has identified the risk factors, causes, and effective interventions for preventing and reducing the rates of CAUTI. Systematic reviews of research demonstrated there are several risk factors for CAUTIs, including older age, female sex, and diabetes[21]; specifically, the most significant risk factors are use of indwelling urethral catheterization, and the length of time the catheter stays in place.[26] Based on this research, The Center for Disease Control and Prevention (CDC) issued CAUTI prevention guidelines in 2009, with the most recent update in 2017. The CDC recommends inserting urinary catheters only for specific indications, and removing them as soon as they are no longer medically indicated to reduce infections. The guideline also provides evidence-based recommendations for proper techniques for catheter insertion and maintenance, quality improvement, administrative infrastructure, and surveillance.[7]

Integrative and systematic reviews of CAUTI-related studies, costs, and interventions in various settings and across patient populations is used as the basis for implementing the EBP process to determine "best fit/best practices" for preventing and addressing CAUTI within your specific organization and patient population.[21, 23–28] Indeed, the extent to which all interventions from research studies are generalizable to all patients in all settings varies; hence *EBP protocols* may differ depending upon the specific population. Some interventions have been validated and generalizable to large populations, such as not using indwelling urethral catheters solely to manage urinary incontinence for convenience on all hospitalized patients.[6] Other interventions, however, are likely to require individualized risk-benefit analysis to select individual patients or patient groups that potentially may have greater benefit from a given intervention. For example, the use of an antiseptic-coated or antibiotic-impregnated catheter has been shown to have only a minor reduction in infection, yet be more costly and potentially more uncomfortable for patients.[25] Thus, use of antiseptic-coated catheters may not be indicated for all patients, but instead be reserved for particularly high-risk populations, such as those with neurogenic bladders. Research has continued to validate CDC guideline recommendations[23, 27] and several EBP initiatives have been conducted to demonstrate development, implementation, and evaluation of protocols (often nurse-driven protocols) and bundles to address the CAUTI problem.[21, 22, 25–27, 29]

The *quality improvement process* is used to evaluate the effectiveness of the EBP initiative, suggest protocol modifications if needed that may be specific to your health system or patient population, and provide continued surveillance to ensure sustainability of the practice change and improved CAUTI outcomes over time.[30] Deliberate, incremental, progressive changes implemented over time with appropriate educational, administrative, and system infrastructure support can then cultivate a culture transformation to embrace sustainable best practices throughout a health system.[31] Through the appropriate use of research, EBP, and QI processes, improvements are being realized in CAUTI prevention, as well as incidence, severity, and cost reduction, with an impact on overall patient outcomes and healthcare professional cultures.

References

1. Shorr RI, Chandler AM, Mion LC, et al. Effects of an intervention to increase bed alarm use to prevent falls in hospitalized patients: a cluster randomized trial. *Ann Intern Med.* 2012;157(10):692–699.
2. Hartung B, Lalonde M. The use of non-slip socks to prevent falls among hospitalized older adults: a literature review. *Geriatr Nurs (New York, NY).* 2017;38(5):412–416.
3. The Joint Commission. *Comprehensive Accreditation Manual for Hospitals. Restraint and Seclusion Standards.* Oakbrook Terrace, IL: The Joint Commission; 2015.
4. Services. DoHaH. *42 CFR Part 482. Hospital Conditions of Participation: Patients' Rights; Final Rule.* Services CfMM, editor, Washington, DC: Federal Register; 2006.
5. Reinhardt AC, Ray LN. Differentiating quality improvement from research. *Appl Nurs Res.* 2003;16(1):2–8.
6. Lang TA, Altman DG. Basic statistical reporting for articles published in biomedical journals: the "Statistical Analyses and Methods in the Published Literature" or the SAMPL Guidelines. *Int J Nurs Stud.* 2015;52(1):5–9.
7. Lang TA, Secic M. *How to Report Statistics in Medicine: Annotated Guidelines for Authors, Editors, and Reviewers.* Philadelphia, PA: American College of Physicians Press; 2006.
8. Woods JC, Mion LC, Connor JT, et al. Severe agitation among ventilated medical intensive care unit patients: frequency, characteristics and outcomes. *Intensive Care Med.* 2004;30(6):1066–1072.
9. Dang D, Melynk BM, Fineout-Overholt E, Ciliska D, Cullen L, Cvach M. Models to guide implementation and sustainability of evidence-based practice. In: *Evidence-Based Practice in Nursing & Healthcare: A Guide to Best Practice.* 3rd ed. Philadelphia, PA: Wolters Kluwer Health; 2015:274–315.
10. DiCenso A, Ciliska D, Guyatt G. Introduction to evidence-based nursing. In: DiCenso A, Ciliska D, Guyatt G, eds. *Evidence-Based Nursing: A Guide to Clinical Practice.* St. Louis, MO: Elsevier; 2004:3–19.
11. Melynk BM, Fineout-Overholt E. *Evidence-Based Practice in Nursing & Healthcare: A Guide to Best Practice.* 3rd ed. Philadelphia, PA: Wolters Kluwer Health; 2015.
12. Stetler CB. Stetler model. In: Rycroft-Malone J, Bucknall T, eds. *Models and Frameworks for Implementing Evidence-Based Practice: Linking Evidence to Action.* Wiley-Blackwell and SIgma Theta Tau International; 2010:51–81. 1st ed. Evidence-Based Practice Series.
13. Titler MG, Kleiber C, Steelman VJ, et al. The Iowa Model of evidence-based practice to promote quality care. *Crit Care Nurs Clin North Am.* 2001;13(4):497–509.
14. Larrabee JH. *Nurse to Nurse: Evidence-Based Practice.* New York: McGraw-Hill; 2009.
15. Melnyk BM, Fineout-Overholt E, Giggleman M, Choy K. A test of the ARCC(c) model improves implementation of evidence-based practice, healthcare culture, and patient outcomes. *Worldviews Evid Based Nurs.* 2017;14(1):5–9.

16. Rycroft-Malone J. The PARIHS framework—a framework for guiding the implementation of evidence-based practice. *J Nurs Care Qual.* 2004;19(4):297–304.
17. Melnyk BM. Models to guide the implementation and sustainability of evidence-based practice: a call to action for further use and research. *Worldviews Evid Based Nurs.* 2017;14 (4):255–256.
18. Melnyk BM, Fineout-Overholt E, Stillwell SB, Williamson KM. Evidence-based practice: step by step: the seven steps of evidence-based practice. *Am J Nurs.* 2010;110 (1):51–53.
19. Milat AJ, Li B. Narrative review of frameworks for translating research evidence into policy and practice. *Public Health Res Pract.* 2017;27(1).
20. Rycroft-Malone J, Harvey G, Seers K, Kitson A, McCormack B, Titchen A. An exploration of the factors that influence the implementation of evidence into practice. *J Clin Nurs.* 2004;13(8):913–924.
20a. Gallagher-Ford L. EBP, Research Innovation, and QI Alignment Model. In: *Personal communication.* 2014 Accessed 28 November 2018.
21. Durant DJ. Nurse-driven protocols and the prevention of catheter-associated urinary tract infections: a systematic review. *Am J Infect Control.* 2017;45(12):1331–1341.
22. Galiczewski JM, Shurpin KM. An intervention to improve the catheter associated urinary tract infection rate in a medical intensive care unit: direct observation of catheter insertion procedure. *Intensive Crit Care Nurs.* 2017;40:26–34.
23. Lam T, Omar MI, Fisher E, Gillies K, MacLennan S. Types of indwelling urethral catheters for short-term catheterisation in hospitalised adults. *Cochrane Database Syst Rev.* 2014;(9):1–4 Art. No. CD004013.
24. Hollenbeak CS, Schilling AL. The attributable cost of catheter-associated urinary tract infections in the United States: a systematic review. *Am J Infect Control.* 2018;46 (7):751–757.
25. Cooper F, Alexander CE, Sinha S, Omar MI. Policies for replacing long-term indwelling urinary catheters in adults. *Cochrane Database Syst Rev.* 2016;(7):1–43 Art. No. CD011115.
26. Gould D, Gaze S, Drey N, Cooper T. Implementing clinical guidelines to prevent catheter-associated urinary tract infections and improve catheter care in nursing homes: systematic review. *Am J Infect Control.* 2017;45(5):471–476.
27. Galiczewski JM. Interventions for the prevention of catheter associated urinary tract infections in intensive care units: an integrative review. *Intensive Crit Care Nurs.* 2016;32:1–11.
28. Fasugba O, Koerner J, Mitchell BG, Gardner A. Systematic review and meta-analysis of the effectiveness of antiseptic agents for metal cleaning in the prevention of catheter-associated urinary tract infections. *J Hosp Infect.* 2017;95(3):233–242.
29. Oman KS, Makic MBF, Fink R, et al. Nurse-directed interventions to reduce catheter-associated urinary tract infections. *Am J Infect Control.* 2012;40(6):548–553.
30. Regagnin DA, da Silva Alves DS, Maria Cavalheiro A, et al. Sustainability of a program for continuous reduction of catheter-associated urinary tract infection. *Am J Infect Control.* 2016;44(6):642–646.
31. Maxwell M, Murphy K, McGettigan M. Changing ICU culture to reduce catheter-associated urinary tract infections. *Can J Infect Control.* 2018;33(1):39–43.

Further reading

32. Gould CV, Umscheid CA, Agarwal RK, Kuntz G, Pegues DA, Healthcare Infection Control Practices Advisory Committee. Guideline for Prevention of Catheter-Associated Urinary Tract Infections (2009). Available from: https://www.cdc.gov/infectioncontrol/guidelines/cauti/; 2017.

CHAPTER FOUR

Financial data in hospitals and health systems

Melissa Kline*, Kelly Seabold†, Molly McNett‡
*Department of Hospital Administration, The MetroHealth System, Cleveland, OH, United States
†Department of Nursing, The MetroHealth System, Cleveland, OH, United States
‡The Helene Fuld Health Trust National Institute for Evidence-Based Practice in Nursing & Healthcare, College of Nursing, The Ohio State University, Columbus, OH, United States

Contents

Hospital reimbursement for provided services

There are a variety of models that impact how hospitals receive payment for services that are provided to patients. In the United States, previous models for health systems included a "fee for service" structure, where health systems were reimbursed for the number of procedures completed, or for the number of patients treated. Under this model, health systems could earn substantially more money by increasing volume of either category. However, the downside of this model is that costs for healthcare were rising faster than the Gross Domestic Product, and the quality of care delivered was beginning to be compromised.[1] The US government, as the largest insurer providing reimbursement on both state and federal levels (Medicare, Medicaid, and military benefits),[2] took note. At the same time, the Institute for Healthcare Improvement issued a "Triple Aim" highlighting the importance of: (1) improving the patient experience; (2) improving the health

Data for Nurses
https://doi.org/10.1016/B978-0-12-816543-0.00004-2

of populations; and (3) decreasing the cost of healthcare to optimize health system quality and performance.[3]

Therefore, in 2010, the US Congress passed the Patient Protection and Affordable Care Act (ACA) to address the rising costs of healthcare and to shift healthcare reimbursement from volume of services to the *value* of services.[4, 5] In order to determine the value of services, several key initiatives then emerged, which established reportable benchmarks and a reimbursement model based on performance. Many of the current benchmarks are reported in Chapter 2. The relationship between these benchmarks and reimbursement structures for value-based care are explained by highlighting important components in recent healthcare reform in the United States.

When the ACA was established in the United States, several reforms were introduced to improve quality and efficiency within the healthcare system and increase public health and prevention services. Two of the primary reforms include the Patient-Centered Medical Home (PCMH) and the Accountable Care Organization (ACO).[6]

Patient-Centered Medical Home

The PCMH focuses on a continuum-based wellness model and is oriented to the whole person.[7] It has five features:

- Patient-centered and emphasizes a partnership between provider, patient, and family;
- Comprehensive care addressing wellness, acute and chronic illness, and end-of-life care;
- Coordinated care across the healthcare system;
- Accessible services; and
- Commitment to safety and quality improvement processes.[6]

The interprofessional team therefore must have shared values, effective communication strategies, and understand the roles of other care professionals and teams, utilizing all of these to meet the goals of the patient.[7] The primary focus of the PCMH centers on care delivery within an individual practice. It involves coordination of care to improve efficiency, access and comprehensive care delivery within a practice. Typically care is overseen by a primary provider, who has ability to refer patients to specialists, and coordinates with those specialty services to optimize quality and efficiency of care provided. By coordinating care under a primary provider while still providing access to specialty services, the program aims to decrease costs and promote positive patient outcomes.

Accountable Care Organizations

In the ACO model, providers form networks to care for a defined population with the aim of decreasing costs and increasing care quality.[8] Essentially ACOs are comprised of a network of PCMHs. Preestablished goals for quality and efficiency are set by government insurers and, if met, the healthcare system receives a financial incentive payment. Conversely, if goals are not met, the system receives a financial penalty.[6] ACOs are meant to encourage providers to change the organization and delivery of care to reduce needless services, prevent unnecessary hospital admissions and readmissions, and meet quality benchmarks.[8] As such, all providers in the ACO have a shared responsibility and vested interest in providing efficient, cost effective, and high quality care.

Specific Reimbursement Programs

Within the overall ACO model, patients are assigned retrospectively to an ACO based on the primary care services they received under prior models, with providers billing as normal and receiving standards fee-for-service payments.[10] A total cost comparison is then done against risk-adjusted target expenditures. For quality measures, a quality score is derived from the average percentage score from each of four domains: patient/caregiver experience, care coordination/patient safety, preventative health, and at-risk population measures.[11] Additional bonuses are issued (or penalties assessed) based on share above the minimum savings threshold and quality performance.[10]

Not all health systems are designated as ACOs or PCMHs. However financial reimbursement for services, regardless of ACO or PCMH status, remains linked to quality outcomes under CMS in the US. Various programs have been established by CMS to financially incentivize hospitals to provide high-quality patient care, such as the Hospital Value-Based Purchasing (VBP) Program, Hospital Readmissions Reduction Program, and the Hospital-Acquired Condition (HAC) Reduction Program.[9]

Hospital VBP Program

The Hospital VBP Program assesses penalties or grants incentives to hospitals based on four domains of care:

- *Clinical Care:* Mortality measures and complication rates
- *Person and Community Engagement:* Hospital satisfaction scores
- *Safety*: Hospital-acquired infection measures and one perinatal care measure
- *Efficiency and Cost Reduction*: Medicare spending per beneficiary

Through the VBP program, individual hospital performance is captured for a set "baseline period" and then compared to the "performance period" for which financial incentives/penalties are assessed. CMS then uses this comparison to determine achievement and improvement in quality and applies a Total Performance Score (TPS) to the hospital. The TPSs for individual hospitals or health systems are compared to the TPS from other hospitals and penalties or incentives are assigned and distributed by CMS. Underperforming organizations will have Medicare Severity Diagnosis-Related Group (MS-DGG) payments reduced up to 2% for each inpatient discharge and increases payments to top-performing hospitals. This redistribution of funding by CMS motivates hospitals and health systems to provide quality and safe clinical care, with a focus on patient satisfaction and efficiency.[9]

Hospital Readmissions Reduction Program

Another program that provides incentives for high quality care is the Hospital Readmissions Reduction Program. This program encourages health systems to focus efforts on reducing unplanned hospital readmissions for six primary diagnoses identified by CMS: acute myocardial infarction (AMI), chronic obstructive pulmonary disease (COPD), heart failure (HF), pneumonia, coronary artery bypass graft (CABG) surgery, and total hip/total knee arthroplasty (THA/TKA). Organizations that maintain low unplanned readmission rates for these diagnoses can avoid financial penalties of up to 3%.[9] Throughout the year, the hospital provides CMS with specific data pertaining to these diagnoses, including, but not limited to, number of patients discharged and readmitted for each diagnosis. CMS then analyzes the data for all contributing organizations and assesses penalties based on the organization's performance as compared to organizations with similar Medicare and Medicaid populations. The organization's excess readmission ratios (ERR) are calculated and compared to the median ERR of the peer group for each diagnosis. When the organization's ERR is higher than the median, a penalty of up to 3% is assessed based on the deviation. Table 1 displays an example of this strategy for key diagnosis codes.

Hospital Acquired Condition Reduction Program

With the switch to pay for performance, healthcare systems are no longer reimbursed for preventable and excessive lengths of in-patient stays and avoidable post-admission complications or Hospital-Acquired Conditions

Table 1 Hospital readmissions reduction program example

	City hospital	Peer group median	Penalty assessed	Base operating MS-DRG payment amount	City hospital payment adjustment factor	% of payment reduction [(10,000 − payment adjustment factor) × 100]	Adjusted payment distributed by medicare	Loss per MS-DRG	# of patients in MS-DRG	Total loss
EER AMI	0.9951	1.0065	No	$4401	0.9826	NA	$4401	$0	801	$0
EER COPD	1.0812	1.0103	Yes	$5562	0.9826	1.74%	$$5465	$97	641	$62,177
EER HF	1.038	1.0233	Yes	$6026	0.9826	1.74%	$5921	$105	1738	$182,490
EER pneumonia	1.0186	1.0192	Yes	$5707	0.9826	NA	$5707	$0	1046	$0
EER CABG	1.0178	0.9969	Yes	$36,836	0.9826	1.74%	$36,195	$641	442	$283,322
EER THA/ TKA	0.9791	0.9814	No	$11,837	0.9826	1.74%	$11,631	$206	977	$201,262
										$729,251

AMI, acute myocardial infarction; *CABG*, coronary artery bypass surgery; *COPD*, chronic obstructive pulmonary disease; *ERR*, excess readmissions ratio; *HF*, heart failure; *MS-DRG*, medicare severity diagnosis-related group; *THA*, total hip arthroplasty; *TKA*, total hip arthroplasty.

(HACs).[12] For most admissions, the system receives a set payment based on a patient's Diagnosis-Related Group (DRG), a patient classification scheme that relates the type of patients treated with the hospital's incurred costs.[12] DRG payments are established through a complex system that takes a multitude of factors into consideration, including geographical labor costs, volume of low-income patients, and teaching status. Because HACs increase the length of stay and resources utilized, CMS will no longer pay for care related to them and most commercial insurers have followed suit. Table 2 lists HAC conditions that CMS will no longer reimburse.[13] A case example is provided to highlight how these HAC impact financial reimbursement for hospitals and health systems.

Case example

Catheter-associated urinary tract infections (CAUTIs) are one of the more frequent hospital acquired conditions (HACs). Costs to treat a CAUTI range from $2800–$3800 per patient. If a multidrug-resistant organism is the causative agent, costs can be even higher. For every patient that develops a CAUTI, typical length of stay is increased by 0.4 to 2 days. This increase in hospital days also increases patient risk for development of other HAC, including *Clostridium difficile*, which can develop as a result of the antibiotics used to treat the CAUTI. Because these additional costs are not reimbursed, a health system will lose substantial amounts of money for each CAUTI acquired, along with subjecting patients to increased risks, and not having that bed space available to treat other patients. For even just 50 simple CAUTIs per year, a healthcare system loses $140,000 to $190,000 just to treat the condition. These costs do not include the additional costs of caregivers, the loss of revenue from not admitting a different patient into the bed, and any other HACs the patients may develop during their extended stay. Therefore, health system efforts to ensure quality care delivery and minimize HAC result in improved patient care and satisfaction, decreased costs, and increased hospital revenue.

Financial metrics that impact the bottom line

Several measures may impact the ability of a hospital to generate revenue or obtain reimbursement for provided services. As the hospital develops a budget, a model is developed that takes into account the revenue (payment) generated by each patient (unit of service) and the expenses

Table 2 Hospital-acquired conditions not reimbursed.

- Foreign object retained after surgery
- Air embolism
- Blood incompatibility
- Stage III and IV pressure ulcers
- Falls and trauma
  - Fractures
  - Dislocations
  - Intracranial injuries
  - Crushing injuries
  - Burn
  - Other injuries
- Manifestations of poor glycemic control
  - Diabetic ketoacidosis
  - Nonketotic hyperosmolar coma
    - Hypoglycemic coma
    - Secondary diabetes with ketoacidosis
    - Secondary diabetes with hyperosmolarity
- Catheter-associated urinary tract infection (CAUTI)
- Vascular catheter-associated infection
- Surgical site infection, mediastinitis, following coronary artery bypass graft (CABG):
- Surgical site infection following bariatric surgery for obesity
  - Laparoscopic gastric bypass
  - Gastroenterostomy
  - Laparoscopic gastric restrictive surgery
- Surgical site infection following certain orthopedic procedures
  - Spine
  - Neck
  - Shoulder
  - Elbow
- Surgical site infection following cardiac implantable electronic device (CIED)
- Deep vein thrombosis (DVT)/pulmonary embolism (PE) following certain orthopedic procedures:
  - Total knee replacement
  - Hip replacement
- Iatrogenic pneumothorax with venous catheterization

associated with each patient. This unit of service can also be further defined as patient days, patient minutes, number of visits, or number of procedures, for example. Expenses consist of the cost of caregivers (salaries and benefits), and supplies used to render care to a patient. The cost of caregivers is often described as hours per patient day (HPPD).[14] Included in this calculation is the total number of worked hours in a unit in a 24-hour period.[15] It is important to note, however, that for comparison to other units or hospitals, HPPD may only be hours of care provided by registered nurses, licensed vocational/practical nurse, and unlicensed assistive personnel[16] and may not include all care expenses for the unit, such as a manager and unit secretary.

Oftentimes, HPPD is presented as two distinct measures: total nursing HPPD and Registered Nurse (RN) HPPD. Total nursing HPPD includes direct care provided by RNs, licensed vocational/practical nurse, and unlicensed assistive personnel, while RN HPPD is reflective of RNs alone.[16] HPPD is calculated by dividing the number of direct care hours provided by the number of patient days in the given reporting period, typically 1 month. For example, in a high acuity unit with a typical RN to patient ratio of 1:2, the monthly RN HPPD would be 12.0 if the unit maintained 1 RN for every 2 patients around the clock. However, staffing and patient acuity rarely remain constant and nurses must continuously adjust staffing levels to meet the needs of the patients. Therefore, when the acuity level drops on the unit and the nurse is able to safely care for three patients around the clock, the HPPD would drop to 8.0.

When analyzing HPPD data, it is essential to look at clinical outcomes, such as patient falls, hospital-acquired pressure injuries, and hospital-acquired infections, during the reporting period. These clinical outcomes are often impacted by the level of staffing on a unit and it is essential for nurses to evaluate these data in conjunction with HPPD. While flexing staff down can result in lower personnel costs, the occurrence of a hospital-acquired infection will result in significant increased costs to the organization. These costs often may be much greater than the cost of appropriate staffing. Nurses who possess a strong understanding of HPPD can use these data to balance staffing needs and clinical outcomes to advocate for safe and fiscally responsible staffing changes. The following case examples highlight key concepts related to HPPD.

Case examples

Essentially, when revenue of a health system exceeds expenses, the health system makes money. However, if costs exceed revenue, the health system loses money. Consider the following scenarios using this example: A hospital unit's budget is set to provide care for 20 patients, with an average length of stay (LOS) of 3 days, 8 HPPD and reimbursement payment for a specific disease condition of $40,000.

Scenario 1—One patient on this unit has a length of stay of 4 days, which is one extra day beyond what was budgeted for in initial estimates. This has also now resulted in 4 days × 8 HPPD or 32 hours of care/expenses, instead of the original 24 hours of care. Thus, the cost of care for this patient is now higher than expected and budgeted for initially. In addition, the extended length of stay for this patient has also not allowed another new admission to arrive. Therefore, the hospital will lose money on this patient.

Scenario 2—Another patient admitted to this same unit has a LOS of 2 days. This has required only 2 days × 8 HPPD or 16 hours of care/expenses. Therefore, this patient's cost of care was less than originally budgeted for, and also allows for a new patient to be admitted sooner than expected. Thus, hospital will make money on this patient.

When examining HPPD, considerations are also included for unit productivity, which takes into account the staffing resources required to care for patients, and the role of a fluctuating census. Consider the following example:

- A unit is budgeted to have 20 patients and use 8 HPPD per patient or 160 hours of care. This is considered the target or goal of 100%.

On day 1, the unit has only 16 patients but doesn't decrease the number of caregivers from when they have 20 patients. The actual worked hours is 160, but their expected worked hours was 128 (16 patients × 8 HPPD). Thus, their productivity is 128/160 or 80%. This will have a negative effect on finances because the unit has used more staff (cost) than they should have (based on the budget) to care for their census.

Between days 1 and 2, the unit receives several admissions. On day 2, the census is now 22. The unit hasn't received any additional help and has their "normal" 160 hours of care. In this scenario, the unit should have 176 hours of care (22 patients × 8 HPPD), but they are still functioning with original staff estimates. In this scenario, their productivity rises 176/160 or 110%. In a budgetary sense, this scenario will have saved money, as the health system only paid for the 160 hours of care, but administered care for 22 patients instead of 20.

Questions for Consideration: What are potential negative effects of a positive revenue using productivity based on HPPD as the sole metric? What are ramifications of this model from the patient perspective? How could understaffing, which produces net revenue for a health system, result in a net deficit over time?

Resources needed to provide care

Specifically within nursing, it is imperative to have the resources needed to provide care, yet ensure alignment with hospital budgets related to staffing. Unfortunately, it has been difficult to quantify exactly how many nurses and nursing support help are needed to care for a population. One would think that more nurses and staff equal better patient care, but research about nurse staffing and the association with patient outcomes has been somewhat inconclusive.[17] When determining the budgeted HPPD, it is important to incorporate evidence and data into decision making. Some factors to consider include[17, 18]:

- Unit budgeted census: Number of admissions and discharges, intensity of care provided, geography of the unit, standards of care
- Patient-specific characteristics: Communication skills, severity of diagnosis, procedures, time needed for education
- Nondirect patient care and nonworked hours: including education and training of staff, paid time off, time away from the bedside for meetings or projects
- Skill mix: Registered nurses, licensed practical nurses, unlicensed staff, manager
- Quality of Care: Rates and incidence of nursing-sensitive HACs, including pressure injuries, falls, catheter associated infections
- Current standards: Requirements from private accreditation agencies, government entities, professional organizations, evidence-based guidelines

These factors should be considered when making staffing decisions that can impact the quality of care delivered on a specific unit. Ultimately these decisions not only impact clinical care, but also patient satisfaction, safety, and the ability of the organization to generate revenue and avoid financial penalties.

Conclusion

Healthcare reimbursement will continue to focus on high-quality care but with lower reimbursements. Healthcare systems are continually looking for ways to decrease costs, yet not compromise on quality of care, safety, or patient satisfaction. It is imperative for nurse leaders to be data driven when making decisions about resources required to demonstrate the value of nursing care and support the overall mission of the health system.

References

1. Yang YT, Nicholas LM. Obesity and health system reform: private vs public responsibility. *J Law Med Ethics*. 2011;39(3):380–386.
2. Fan N. Rising healthcare costs and the increasing role of government in medicine. *Del Med J*. 2014;86(5):137–138.
3. U.S. Senate. *The Patient Protection and Affordable Care Act: Detailed Summary*. Retrieved from: https://www.dpc.senate.gov/healthreformbill/healthbill04.pdf; n.d.
4. Berwick DM, Nolan TW, Whittington J. The triple aim: care, health, and cost. *Health Aff*. 2008;27(3):759–769.
5. Dempsey C, Reilly B, Buhlman N. Improving the patient experience: real-world strategies for engaging nurses. *J Nurs Adm*. 2014;44(3):142–151.
6. Vincent D, Reed PG. Affordable Care Act: overview and implications for advanced nursing. *Nurs Sci Q*. 2014;27(3):254–259.
7. Martin DM. Interprofessional teams in the patient-centered medical home. *Nurs Adm Q*. 2014;38(3):214–220.
8. National Health Policy Forum. *Pioneer ACOs: The Frontier of Delivery System Reform?* Retrieved from: http://www.nhpf.org/library/forum-sessions/FS_06-13-14_PioneerACOSs.pdf; 2014.
9. Centers for Medicare & Medicaid Services. *Hospital Readmissions Reduction Program (HRRP)*. Retrieved January 21, 2019 from: https://www.cms.gov/medicare/medicare-fee-for-service-payment/acuteinpatientpps/readmissions-reduction-program.html; 2019.
10. Sweeney-Platt J. *How Does an ACO Work?* Retrieved December 31, 2018 from: https://www.advisory.com/-/media/Advisory-com/Research/PEC/Events/Webconference/2014/How-Does-an-ACO-Work-052914.pdf; 2014.
11. Centers for Medicare & Medicaid Services. *Medicare Shared Savings Program: Quality Measure Benchmarks for the 2018 and 2019 Reporting Years*. Retrieved December 31, 2018 from: https://www.cms.gov/Medicare/Medicare-Fee-for-Service-Payment/sharedsavingsprogram/Downloads/2018-and-2019-quality-benchmarks-guidance.pdf; 2017.
12. Centers for Medicare & Medicaid Services. *Design and Development of the Diagnosis Related Group (DRG)*. Retrieved December 1, 2018 from: https://www.cms.gov/ICD10Manual/version34-fullcode-cms/fullcode_cms/Design_and_development_of_the_Diagnosis_Related_Group_(DRGs)_PBL-038.pdf; 2016.
13. Centers for Medicare & Medicaid Services. *Hospital-Acquired Conditions*. Retrieved December 1, 2018 from: https://www.cms.gov/Medicare/Medicare-Fee-for-Service-Payment/HospitalAcqCond/Hospital-Acquired_Conditions.html; 2018.
14. Hunt PS. Developing a staffing plan to meet inpatient unit needs. *Nurs Manage*. 2018;49(5):25–31.
15. Fike GC, Smith-Stiner M. Hours per patient day: understanding this key measure of productivity. *Am Nurse Today*. 2016;11(4).
16. Press Ganey Associates, Inc. *Guidelines for Data Collection and Submission on Nursing Care Hours (NCH) Indicator*. Overland Park, KS: Author; 2018.
17. He J, Staggs VS, Bergquist-Beringer S, Dunton N. Nurse staffing and patient outcomes: a longitudinal study on trend and seasonality. *BMC Nurs*. 2016;15:60–64.
18. Ohio Revised Code. Hospitals: Duties of nursing care committee. In: *Ohio Revised Code: Title XXXVII—Health-Safety-Morals*; 2008. [chapter 3727] Retrieved from: http://codes.ohio.gov/orc/3727.53.

CHAPTER FIVE

Evaluating data to guide care delivery: Quality improvement methods and implementation science

Sarah Livesay*, Mary Zonsius*, Molly McNett†
*Adult Health and Gerontological Nursing, Rush University, College of Nursing, Chicago, IL, United States
†The Helene Fuld Health Trust National Institute for Evidence-Based Practice in Nursing & Healthcare, College of Nursing, The Ohio State University, Columbus, OH, United States

Contents

Introduction

Quality improvement (QI) has become an essential competency and a daily skill for nurse leaders.[1] Unfortunately, few nurses are taught about QI

Data for Nurses
https://doi.org/10.1016/B978-0-12-816543-0.00005-4

formally during orientation efforts or as part of ongoing professional development and training within hospitals and health systems. While the American Association of Colleges of Nursing (AACN) includes leading QI initiatives as an essential competency of Doctor of Nursing Practice (DNP)-prepared nurses,[2] nurses at all levels of educational preparation are regularly participating in and in some cases, leading QI initiatives. This chapter will provide a brief history of the evolution of QI in healthcare, concurrent with the development of an overarching model for translational science. Inherent in translational science approaches are methods to evaluate uptake of research findings into routine clinical practice, including implementation science (IS) and QI methodologies. The relationship between research, evidence-based practice (EBP), QI, and IS will be discussed, with a focus on applications to nursing practice.

History and evolution of the QI movement

As discussed in Chapter 1, the modern healthcare QI movement in the United States was founded by two seminal documents in 1999 and 2001: *To Err is Human* and *Crossing the Quality Chasm*.[3, 4] Written by the Institute of Medicine (now referred to as the National Academies of Medicine), these documents outline the safety and quality concerns of healthcare delivered in the 1980s and 1990s, suggesting that up to 98,000 people die every year due to medical error. While these numbers were initially challenged by the healthcare community, data in later years suggest this number is much higher.[5] *Crossing the Quality Chasm* was published several years later and offered a critique of the current US healthcare system while envisioning a future system predicated on safety, quality, and cost effectiveness.[3] The paper suggested that in the current healthcare system, safety and quality of care is an individual responsibility, and improvement is reactive. However, in the proposed future healthcare system, safety and quality should be a system responsibility and programs should be surveilling for opportunities to improve safety and quality, rather than simply reacting to adverse events. A series of subsequent publications further emphasized the importance of improving safety and quality in healthcare, including the Institute for Healthcare Improvement's Triple Aim, which charges healthcare systems to[1] improve the patient's experience of care (including quality and satisfaction),[2] improve the health of populations, and[3] reduce the per capita cost of healthcare.[6]

Collectively, these initial reports ignited a substantial shift in the perspective of QI across health systems. Healthcare quality was now inextricably linked to the system providing care. As discussed in Chapter 4, health systems' measures of quality were directly impacting financial reimbursements and penalty fees; hence, departments of quality management rapidly developed and grew within health systems as a mechanism to direct and oversee quality of care. During this time, the Centers for Medicare and Medicaid within the United States launched the inpatient quality reporting program commonly referred to as "core measures," and later launched a parallel outpatient quality reporting program.[7] As discussed in Chapter 2, additional quality measures also became increasingly prevalent as disease-specific certifications (i.e., national certifications for stroke, heart failure, cardiac specialty centers, etc.) grew in popularity and required surveillance and reporting of key quality indicators and benchmarks within specific patient populations. Numerous healthcare stakeholders now require quality or ongoing reporting of performance data, mandating that health systems have a robust infrastructure for QI management. A key component of that robust infrastructure includes members of the healthcare team, including nurses, actively leading and participating in QI to ensure efficient delivery of evidence-based practices to optimize outcomes for patients.

The process of QI in healthcare settings often falls under the umbrella of both IS and EBP, as QI initiatives in healthcare are typically aimed at addressing implementation and assimilation gaps between research and clinical practice. Ultimately these efforts are key components in translational science, which includes development of knowledge "from bench to bedside." The National Institutes of Health describes translational science as a spectrum of knowledge development that includes the following components (NIH)[8]:

- Basic bench and/or biological research;
- Preclinical research using animal or cell models;
- Clinical research investigating impact on patients;
- Clinical implementation, which examines how well findings can be incorporated into routine clinical practice;
- Public health evaluation and surveillance to determine how knowledge translates into improved population health.

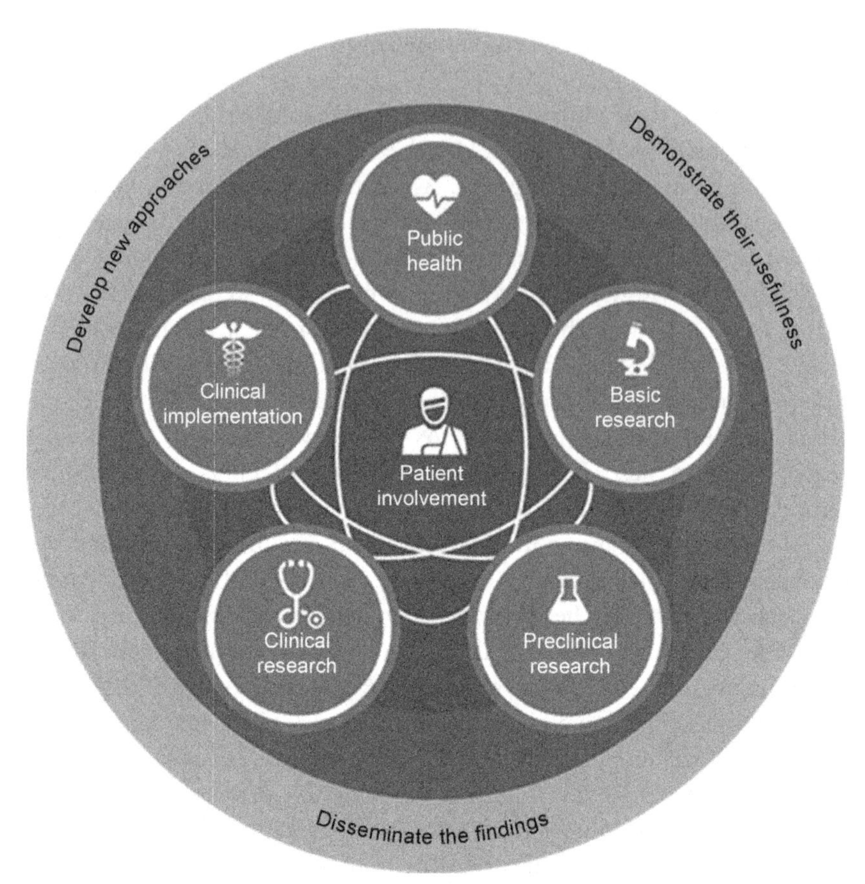

Fig. 1 Translational science model.

It is important to note the process is not always linear; rather all stages can be intertwined and include development of knowledge to advance other stages. Fig. 1 highlights the proposed model for translational science from the National Center for Advancing Translational Science.[8]

Table 1 displays a description of the different types of investigations typically encompassed in the key components of the model. It is important to note that these stages do not occur in isolation, and there may be overlap between methodologies used to advance the science and application to clinical practices. In addition, the process is not always linear, nor does it ultimately end solely with applications to public health settings. Rather, the process can be iterative and cyclical based on findings from the various stages and methodologies.

Table 1 Examples of components within translational science spectrum[8]

Translational science descriptor	Translational science stage	Examples of common approaches
Basic research	T0	Preclinical bench research Biological research Animal studies
Preclinical research	T1	Studies investigating translation to humans Proof of concept studies Early feasibility and efficacy trials Phase I clinical trials on healthy volunteers
Clinical research	T2	Studies evaluating translation to patients with a specific condition Data from these trials often used to generate EBP recommendations Phase II-III clinical trials
Clinical implementation	T3	Investigates how to integrate into routine clinical practice Implementation science, QI science, QI, and evidence-based practice methodologies Comparative effectiveness research
Public health	T4	Public health and population health research

Implementation science

Implementation science focuses on methods used to identify and promote the systematic and consistent uptake of research and other data sources to improve effectiveness of healthcare delivery and minimize the research to practice gap.[9, 10] The focus of IS is to identify what is required to actually implement knowledge about what promotes positive patient outcomes into routine practice.[9] Implementation lies on the spectrum of knowledge diffusion-dissemination-implementation, where diffusion refers to the generation of new knowledge, dissemination includes sharing this knowledge with target audiences, and implementation centers on how to integrate this new knowledge into a specific setting.[11] Therefore, the science of implementation centers on eliminating factors that impede knowledge utilization, while enhancing factors that promote uptake of best evidence into practice. The innovations and knowledge gained from research and other data sources is critically important for advancing health; however, equally important is the ability of healthcare providers to actually be able to implement these strategies into their daily practice and produce the same positive results. IS therefore focuses on how to institute these best practices into everyday care of patients within health systems.

Relationship between implementation science and quality improvement

Because IS focuses on optimizing uptake of best practices into routine clinical care, it can include some overlap with QI methodologies. For example, QI methods may be used to identify the best approach to implement a best practice into routine care using rapid cycle change, or a critical analysis of causative factors for research uptake and utilization. It is important to note, however, that there are key differences between the two strategies. Table 2 highlights these fundamental differences.

As illustrated in Table 2, the triggers are often different, as QI begins as a result of a problem or issue in a care area, while IS begins with an evidence-based practice. Both have methodological rigor and overall aim to improve care delivery. However, the methodologies are different, and short-term gains and direct outcome goals are different. QI is often focused on improving a process to positively impact a patient or population outcomes. In contrast, IS focuses on identifying effective methods to incorporate routine use of an evidence-based practice into daily care delivery. While the outcome ultimately in IS overall is to improve patient outcomes, the direct goal or immediate gain is to identify effective strategies for the implementation of best practices into routine clinical care.[9]

Table 2 Key differences between QI and IS[9]

	Quality improvement	Implementation science
Origin/trigger	Specific problem in care area	An evidence-based practice
Purpose	Determine "what is"	Determine "how to"
Method	Design/pilot strategies to address the problem. Use structured methodology, such as Lean, Six Sigma, or Plan-Do-Study-Act methods.	Design strategies to address gap in the healthcare system or care area that is preventing use of the EBP. Can include some QI methodologies if applicable.
Direct goal	Improve patient outcomes and/or care processes.	Identify effective strategies for implementation of the EBP into routine practice.

IS also differs from research, yet shares some common characteristics. The overarching goal of research is to generate new knowledge; the process typically involves planned interventions and measuring impact on outcomes under controlled conditions. In contrast, IS seeks to identify the optimal method for consistent and sustained uptake of the EBP from those research studies—and how to make that work as part of the usual care in the healthcare setting. IS includes the concept of "naturalistic variability"—or identification of contextual factors that can impact day-to-day utilization of the best evidence when making decisions about patient care, and evaluates how the best practice can be implemented when not controlling for key variables.[10]

IS also differs from traditional research. Table 3 highlights key differences. Implementation studies differ from traditional research studies in several ways. Both approaches employ different purposes, methods, and outcomes. As displayed in Table 3, research typically focuses on generating new knowledge to optimize patient outcomes, while IS focuses on generating information about optimal techniques to incorporate EBP into every day practice. The hypothesis and unit of observation for a traditional study is often centered on patient outcomes, while IS centers on the healthcare

Table 3 Differences between traditional research and implementation research

	Traditional research study	Implementation study
Purpose	Generate new knowledge on what impacts patient outcomes	Generate new knowledge on what works to implement that research or EBP into daily practice
Hypothesis	Focuses on a patient outcome	Focuses on the ability of the team to use the EBP
Intervention	Patient, provider, or system-level functions that impact the outcome. Specific variables are included a priori and measured under controlled conditions	Methods or techniques to institute change and implement strategies (bundle)
Unit of observation	Patient	Providers, caregivers
Outcome	Effect of intervention on patient outcomes	Rate and quality of use of the EBP
Priority	Internal validity	External validity

environment, specifically the individuals providing care. Variables in traditional studies include patient and clinical factors impacting the outcome under investigating, while IS variables center on contextual and behavioral factors that impact uptake and routine integration of the new knowledge that was generated from the research studies. There are key considerations in traditional research to strengthen and ensure internal validity of the study. In contrast, IS centers on external validity and the ability to share information about effective approaches for EBP uptake to wider audience to facilitate continued use of EBP to guide practice.[9]

Within the United States, but also on an international level, there is an important and renewed focus on quality of care, which includes how to effectively integrate findings from research into practice. Contextual factors can influence how a best practice is utilized with different health systems, and can include available resources, infrastructure, knowledge and educational preparation of staff, equipment, and characteristics of patient populations served by the health system. IS has therefore emerged as a critical concept to determine the best way to implement identified best practices into routine care, with considerations for local and contextual factors present within a specific health system(s).

IS theories, models, and frameworks

As with research and QI, there are several theories, frameworks, and models to guide IS. *Theories* generally refer to relationships between key constructs, concepts, and/or variables and explanations that predict specific events based on these relationships.[12] An example of a theory used in IS is Roger's Theory of Diffusion of Innovations, which provides an explanation for how new ideas or practices gain momentum and spread (or diffuses) over time through a specific group of individuals.[13]

In contrast, *models* used in IS are commonly more narrow in scope and are descriptive instead of explanatory when outlining key components.[12] Examples of models used in IS include the Stetler Model, the Iowa Model, and the ACE Star Model of Knowledge Transformation, which are described in Chapter 3. Similar to models, *frameworks* do not posit to explain causal relationships between concepts; rather, frameworks present a general outline or structure of categories, and aim to fit phenomena or variables into each category.[12] The Knowledge to Action Framework (Fig. 2) is an example that can be utilized in IS.[14]

Knowledge to Action Framework: Key components
Assess the problem
Adapt to local context
Identify barriers and facilitators
Determine implementation strategies
Process surveillance and evaluation of outcomes
Sustainability considerations

Fig. 2 Knowledge to Action Framework.

CFIR framework: Key components
Identify the intervention
Assess the inner setting when intervention will take place
Assess the outer context surrounding implementation
Identify individuals involved with implementation
Outline the process of how the intervention will be performed

Fig. 3 CFIR framework.

Similarly, many IS groups have relied upon Consolidated Framework for Implementation Research (CFIR).[15] This approach allows for flexibility with implementation, and components do not have to occur in a stepwise fashion. Key components include considerations for individuals involved in implementing the science, and are displayed in Fig. 3.

Implementation strategies

Overall there are commonalities among the various models and frameworks used in IS. A summary of expert recommendations for implementation change projects identifies seven overall categories of strategic approaches to IS. Categories include a component of evaluation (auditing and feedback), interactive assistance (facilitation activities), adaptations of the intervention to control for local contextual factors (tailoring strategies), establishing stakeholder relationships (i.e., identifying champions, mentors, key players), training and education, methods to support clinicians, and engagement of consumers and students.[16] Overall it is important to assess for readiness, including identification of facilitators and barriers, embedding mechanisms for feedback and evaluation processes, and in some cases, theories or models of behavior change techniques.[17] A careful consideration of strategies at the unit level, but also at the organizational level is necessary to ensure sustainability over time.

Bundle approach

Use of implementation bundles is often one component among the most effective strategies for uptake and integration of best practices. Bundles should include identification of what the intervention is that needs to be incorporated into practice (the what), followed by a systematic strategy for how to integrate the interventions into daily practice (the how), mobilization of a team to perform the work of executing the intervention (the who), and determining the best location for initial integration of the intervention (the where).[9]

Team approach

Mobilization of implementation teams is an effective approach to address the "who" of implementation science. Similar to the creation of research teams to successfully conduct a study, implementation teams within a health system critically analyze contextual factors that can impede or promote uptake of best evidence into routine practice. Collectively these team members work to minimize barriers for integration, and establish the infrastructure for successful data utilization across care settings.[9] Formation of implementation teams is often effective at efficiently integrating best practices into routine care. PDSA cycles, while typically reserved primarily for QI initiatives, can sometimes be used by the team to evaluate uptake strategies in the short term.

Testing approach

Other methods employed by implementation teams or as part of an implementation bundle can include pilot testing and usability testing to evaluate feasibility of the intervention in local settings. Pilot testing typically involves integrating the intervention within one unit or location over a defined period of time to identify obstacles and facilitators to knowledge uptake and routine integration. Modifications are made to the intervention strategies, and then employed across larger groups or care areas. In contrast, usability testing typically involves four to five iterations of testing across a few sites to allow for continued modifications prior to widespread implementation.[10] Findings from these testing approaches, and observations and experiences of the implementation teams will elucidate key barriers and facilitators for change. These factors may be present at the practice level and easily address, or at the system level, requiring collaboration with executive teams to promote EBP uptake across the health system.

Case example

Multiple trials have evaluated best practices to improve outcomes for patients presenting with acute ischemic stroke (AIS). Bench science has identified possible agents to achieve thrombolysis, and Phase I-III trials have investigated drug development, safety, efficacy, and dosing of certain medications to treat patients who present with symptoms of AIS and meet specific clinical criteria. One medication is alteplase, which was demonstrated in clinical trials to be an effective agent in achieving thrombolysis during AIS. As a result of these trials, clinical practice guidelines from the American Stroke Association (ASA) recommend administration of alteplase within 60 minutes of presentation to the emergency department (ED) among qualifying patients.[18] Recognizing this evidence-based practice (EBP) recommendation, a large, tertiary health system applies the Consolidated Framework for Implementation Research (CFIR) to implement the new recommendation into routine clinical practice. As part of this Implementation Science (IS) approach, a designated team **identified the intervention**, which included new administration times for the medication. They then **assessed the inner setting** where the intervention would take place, which included the emergency department. They **also assessed the outer context** of coordination with pharmacy, radiology, emergency medical services, and the critical care units. It was also necessary to **identify individuals involved with the implementation**, which included ED, pharmacy, radiology, and critical care unit interdisciplinary staff. Finally, the group **outlined the process of the new intervention** in their policies, procedures, and ordersets. As a result, the change was implemented, staff were educated, and the health system expected the reportable benchmarks of time to administration to begin to align with current best practice recommendations.

As highlighted in the case study, clinical research contributes to the science-to-practice gap by generating the evidence that is used in EBP. IS then aims to translate these best practices from research into everyday clinical practice. IS is defined as the study of methods to improve the uptake, implementation, and translation of research into routine and common practice.[19] However, a third gap in translation of evidence into routine practice is often present. A health system may have adopted a new practice based on research evidence, and have worked to incorporate it into daily clinical practice. However, this new practice may not be effectively utilized or sustained, and the expected improvements in patient outcomes are not occurring. Thus, QI initiatives may be incorporated by healthcare teams to address the implementation and assimilation gap in order to identify the best way to routinely implement this best practice to consistently achieve the desired results.

Quality improvement methods

QI methodologies are often described in relation to the emerging field of improvement science. While similar to implementation science, improvement is different in that it aims to identify how and why change occurs. Improvement science has been described by Marshall, Pronovost, and Dixon-Woods (2013)[20] as an approach that "… draws on, and aims to contribute to, clear and explicit theories of how change happens. A major component of improvement science focuses on the design, deployment, and assessment of complex, multifaceted interventions." (p. 419). Thus, QI methodologies may be used to identify optimal design or implementation techniques that may result in facilitating change within an improvement science model.

Quality improvement taxonomy

The language and terms surrounding QI have been modified over the past 20 years as the field has evolved, and these changes are relevant to understanding QI methodologies that guide data management. Because much of the QI infrastructure in healthcare is originally borrowed from the manufacturing industry, terms such as *total quality management* (TQM), *benchmarking*, and even *quality* were originally used to describe QI within healthcare systems. TQM refers to the overall structured approach to quality management. Benchmarking refers to a system that allows a comparison of measurement with other similar facilities or products. Lastly, the term quality is generally associated with the concept of the output or product being free from deficits and variation.[1, 21]

Specifically within healthcare, The US Health Resources and Services Administration (HRSA) defines QI as systematic and continuous actions that led to measurable improvement in healthcare services and health status of targeted patient groups.[19] Individual QI programs and definitions may vary in scope and focus, but all incorporate the following key principles: QI addresses systems and processes, focuses on the patient(s), is team-based, and uses data to guide projects.

Common QI methodologies

A guiding methodology and data are essential for any type of systematic change initiative. As discussed in the previous section, QI often spans both implementation and improvement sciences as each continues to evolve and identify strategies for optimal integration of best evidence into routine and sustained clinical practice. Table 4 provides a pictorial representation of common QI, evidence-based practice (EBP), change and evaluation theories.

While the Donabedian Quality Model, Lean, Six Sigma, and Shewert Cycles are commonly referred to as QI models, many QI projects are a part of larger EBP and/or change initiatives. Integration of evaluation models is

Table 4 Common theories, frameworks, and models

QI	EBP	Change	Evaluation
Donabedian	ACE Star Model	Kotter	CDC Evaluation Model
Lean	ARRC Model	Lewin	Donabedian Quality Model
PSDA/ Shewert	Johns Hopkins EBP	Rogers Diffusion of Innovation	Kellogg Evaluation Framework
Six Sigma—DMAIC	Knowledge to Action		PRECEDE-PROCEED
	Iowa Model		Proctor
	Stetler Model		Rainbow Evaluation
			RE-AIM

ACE, Academic Center for Evidence-Based Practice; *ARCC*, Advancing research and clinical practice through close collaboration; *CDC*, Centers for Disease Control; *DMAIC*, Define, measure, analyze, improve, control; *PDSA*, Plan-Do-Study-Act; *PRECEDE*, Predisposing, Reinforcing and Enabling Constructs in Educational Diagnosis and Evaluation; *PROCEED*, Policy, Regulatory, and Organizational Constructs in Educational and Environmental Development; *RE-AIM*, Reach, Effectiveness, Adoption, Implementation, Maintenance.

critical to evaluate the impact of the change over time, or to develop baseline data when the problem underlying poor quality is not yet known. Therefore, QI methodologies should be considered within the context of overarching models and frameworks.

Donabedian quality model

Avedis Donabedian noted in the 1960s that quality healthcare was composed of three key categories or measures: structure, process, and outcome.[22] The model suggests that all three categories are interrelated and essential to quality care. Structure drives processes, which, in turn, drives outcomes. However, outcomes also are critical in informing process and structure. Therefore, the model can be used in an iterative fashion. The Donabedian Model is particularly useful when developing or evaluating a new or existing program in healthcare. Data are essential to the model, as measurement is a key aspect of the outcomes category. However, certain clinical performance measures may also evaluate structural and process elements for their role on overall program outcomes. Therefore, data may drive specific aspects of each category in this model.

Lean and Six Sigma

Lean is a term coined in the late 1980s to describe manufacturing methodologies aimed at eliminating waste and decreasing deficits during mass production processes.[23] Lean quickly became synonymous with the Toyota Production System (TPS) as the organization adapted the methodology to improve the car-manufacturing process. Because Lean focuses on improving processes and removing waste, quality is thought to improve as a secondary effect. The methodology employs a number of specific tools, including value stream mapping, Ishikawa diagrams, the 5-Whys, Poka Yoke, the 5-S system, Failure Modes Effect Analysis, and Kaizen, among others.[24] Table 5 includes a description of common approaches.

In addition to Lean, Six Sigma, literally translated as 3.4 deficits per million, is a manufacturing methodology developed by a number of large companies. Six Sigma employs a number of quality improvement tools with the goal of decreasing variation in order to improve outcomes. This approach emphasizes the use of statistical tools to analyze and improves performance. Many healthcare organizations have utilized a hybrid model of QI that includes both Lean and Six Sigma approaches to improvement.[24]

Table 5 Common Lean approaches

Six Sigma process	Key focus	Description
Value stream mapping	Modify process to reduce variation and streamline systems.	Shows process flow within three categories: Value enabling, value adding, nonvalue adding. Eliminate nonvalue adding, and decrease delays between other components to increase precision and efficiency.
Ishikawa diagram	Evaluate cause and effect.	Also referred to as "fish bone" diagrams, and are created to show potential causes of a specific outcome or event.
5 Whys	Determine root cause of a situation or process.	Identify the key problem; ask why the problem occurred. Continue to ask "why" up to five times to identify the root of the problem.
Poka-Yoke	Focus on mistake-proofing.	Employees identify causes of human error in processes and work to eliminate or decrease risk of error.
5S system	Organize workspaces to improve efficiency.	Focus on the following components: Seiri (Sort)—Remove any unnecessary items. Seiso (Shine)—Inspect and clean workspace daily. Seiton (Straighten, set)—Organize what is left for optimization. Seiketsu-(Standardize)—Establish standards for sorting and inspections. Shitsuke (Sustain)—Apply the standards across work areas. Focus: Eliminate the nonvalue adding; modify process to reduce variation and streamline systems.
Failure Modes Effect Analysis	Identify and eliminate weak components in early stages of process or development.	Identify failures inherent in early processes, as well as causes and effects to mitigate its occurrence.
Kaizen	Continuous improvement	Engagement of all employees at all times to identify inefficiencies or obstacles that impede processes.

One of the most common Six Sigma methods used in healthcare organizations to evaluate and improve processes is the Define/measure/Analyze/Improve/Control (DMAIC) approach.[24] Within this approach, the five steps for improving processes include:

- Define—identify the problem and key stakeholders
- Measure—collect baseline data
- Analyze—assess the data to identify problems
- Improve—develop and implement solutions to defined problems
- Control—maintain improvement

Shewert/Deming cycle (Plan-Do-Study-Act)

The Plan-Do-Study/Check-Act (PDSA or PDCA) cycle is commonly used for QI in healthcare systems. The cycle was a tool and part of a much larger QI methodology described by Walter A. Shewart and W. Edwards Deming, pioneers of the manufacturing QI movement. The cycle is helpful as it clearly and simply outlines an iterative cycle that can be applied to a variety of problems and settings:

- Plan—Develop goals and processes necessary to improve situation
- Do—Implement improvement activities
- Study/Check—Gather data to determine the impact of the improvement activities
- Act—Evaluate the impact of the improvement activities and determine if any additional action is needed. The decisions made in this phase often inform the next iteration of the PDSA cycle.

In successful QI projects, PDCA cycles are iterative and ongoing, and integrate rapid cycle change.[25] As demonstrated in Fig. 4, often multiple small, rapid-cycle change projects are implemented sequentially to optimize outcomes and meet or exceed targeted benchmarks over time.

It is important to note that within healthcare systems, one QI approach is not always utilized in isolation. Rather, most organizations incorporate a combination of multiple methodologies, borrowing tools from Lean, Six Sigma, and PDSA cycles to optimize results. It is important to note that both Lean and Six Sigma methodologies emphasize a top-down and bottom-up approach to QI, highlighting that quality must be a priority of all levels of an organization and cannot be distilled into just tools that are only applied when convenient. Thus, all members of the healthcare team, including nurses, should be familiar with these approaches for optimizing efficiency, minimizing waste, and positively impacting patient outcomes. Nurses in particular

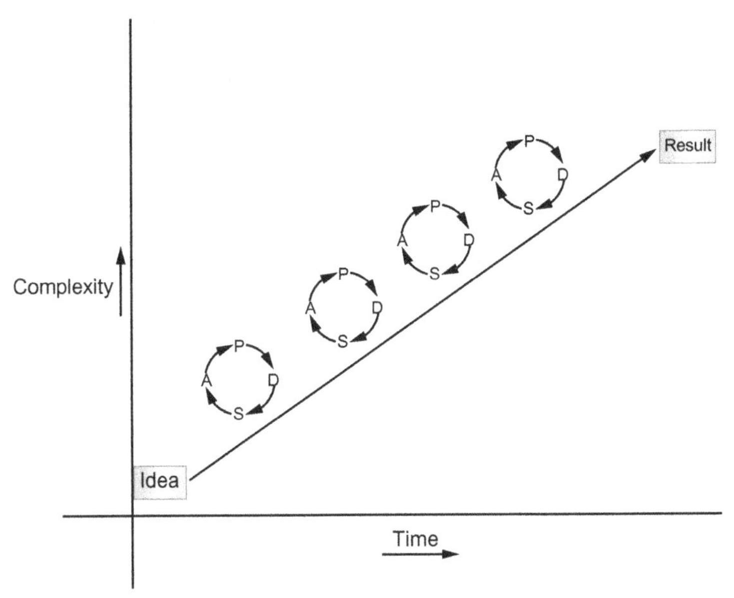

Fig. 4 Rapid-cycle change. Multiple PDSA cycles may be required in building from a change idea to a tangible result. *PDSA*, Plan-Do-Study-Act.

have tremendous perspective on factors influencing care delivery, and therefore have an integral role in mitigating elements using these QI approaches that may impede efficiency or negate outcomes.

Application of methodologies

Healthcare systems in the United States are now required by accrediting agencies to have a robust, system-level quality improvement program, which includes adoption of one or more QI methodologies. Many organizations use a combination of tools and methodologies. Organizations using Lean, Six Sigma, or PDSA methodologies typically provide specialized training for designated members of healthcare teams, and often pair them with dedicated QI specialists to assist with QI projects. Generally speaking, it is best to work within the QI methodologies and structures of the organization and/or program to ensure alignment with hospital benchmarks and initiatives. Regardless of the specific QI methodology used within a health system, it is beneficial to possess a solid understanding of the various methodologies and tools available for QI. The following case examples highlight integration of specific QI methodologies used to improve outcomes for specific populations.

Case example: Six Sigma application

Although the large, tertiary academic medical center presented in the first-case example had implemented the EBP of administration times for alteplase to be within 60 minutes, benchmarking data after 6 months indicated the staff was struggling with door-to-treatment times for administration of intravenous (IV) alteplase to treat patients with acute ischemic stroke. The average administration times were noted by the organization to be worse than those reported by other like hospitals when using a national database to benchmark. Target benchmarks included administration times to be within 60 minutes of hospital arrival. The organization had tried several interventions to improve administration times over the past 6 months, with no significant improvement. They were concerned that the slow administration times was both a threat to patient care because of potentially worsened outcomes, and also a possible vulnerability in their upcoming stroke certification visit. The organization had a robust quality improvement department that primarily employed Six Sigma methodologies. Because of the failure of the stroke program to show improvement with their previous interventions, a trained Six Sigma improvement specialist was assigned to this problem and a formal QI initiative was launched.

Baseline data: The organization administered IV alteplase 42 times in the previous 12 months. The average administration time for the past 12 months was 70.5 minutes, median time 71 minutes, and range was 49–99 minutes.

Six Sigma improvement process

Define	• A QI task force of key stakeholders was formed, a Six Sigma quality consultant was dedicated to guide the project • Alteplase administered in the emergency department at the hospital of interest • Project to span 6 months
Measure	Current state: • IV alteplase administered 42 times in the previous 12 months • Average administration time for the past 12 months: 70.5 minutes, median time: 71 minutes, range: 49–99 minutes. • Half-day interdisciplinary meeting: value stream mapping, fishbone and Pareto analyses conducted to understand the problem
Analyze	Key findings: 1. Significant shift variations with longer times for treatment from 3 p.m. to 7 a.m.

	2. Delays in calling treating neurologist during stroke code, call-back times from treating neurologists **3.** Average time from medication ordered to delivered by pharmacy to bedside 18 minutes
Improve	Action plans: **1.** Rapid response team staff to assist with stroke alerts from 3 p.m. to 7 a.m. **2.** Implement new stroke code process with neurology call at time of stroke alert. Addition of telemedicine equipment to enhance physician availability and treatment. **3.** Pharmacist to respond in-person to stroke alert and mix alteplase
Control	Weekly stroke alert audits to monitor process and progress Checklist to assist with new processes Debrief after each treatment to identify improvements and opportunities Telemedicine training for physician and nursing staff, 24-hour technology support line for first 90 days

After implementation of the new processes and technology, leadership closely monitored the processes and staff debriefs on a weekly basis. Treatment times were improved to an average of 52 minutes within 3 months of implementation. At month 3, it was identified that a direct to computed tomography (CT) scan process may improve treatment times. The action plan was revised to include a quick debrief with emergency medical services in the triage bay on arrival, and transport of the patient straight to CT-scan scan prior to rooming in the emergency department. Measures continued to improve over the next 3 months, with average administration times dropping below 50 minutes.

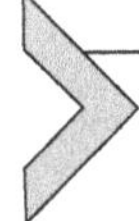

Managing QI data

Back to the basics

In any QI program or project, data management is key to success. Understanding how to view data and what statistical tools should be applied are key skills of the nurse overseeing a QI program. When deciding on appropriate QI tools, it is important to start with the basics. Data collected in QI initiatives may be quantitative or qualitative, and may encompass data sets that are nominal, ordinal, interval, and/or ratio. A key element of developing a data collection and evaluation plan is to understand the type of data collected.

The type of data collected helps to drive the appropriate QI tools used for data analysis and display.

Basic data display may be accomplished through bar graphs, pie charts, run charts, line graphs, and/or Cartesian graphs. Generally speaking, bar graphs are adequate to display numbers or categories that are independent of each other. Pie charts are commonly used to display a total value that is divided into different categories. Line graphs display data points that are related to each other, most often representing data over time. Cartesian graphs (also known as scatter diagrams or scatterplots, *X*-*Y* graphs) include numbers on both the *X* and *Y* axes and demonstrate how a change in one value affects the other. The Figs. 5–8 display examples of common ways to display data.

Basic data display should not be confused with common QI tools used to collect data. For example, tools such as the 5 whys, fishbone analysis, spaghetti graphs, and affinity diagrams are useful QI tools to understand how processes relate to or impact each other. When choosing a tool to use in a QI project, one must contemplate how the data collected from the tool may be displayed for understanding and if the data from the tool lend themselves to any simple statistical analysis.

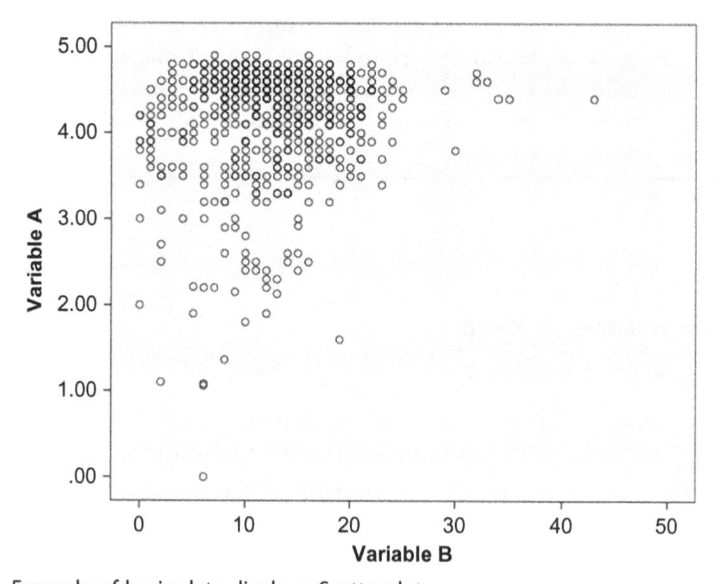

Fig. 5 Example of basic data displays: Scatterplot.

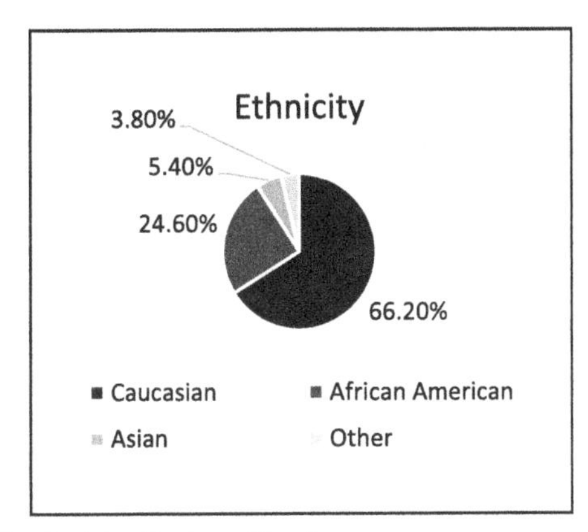

Fig. 6 Example of basic data displays: Pie chart.

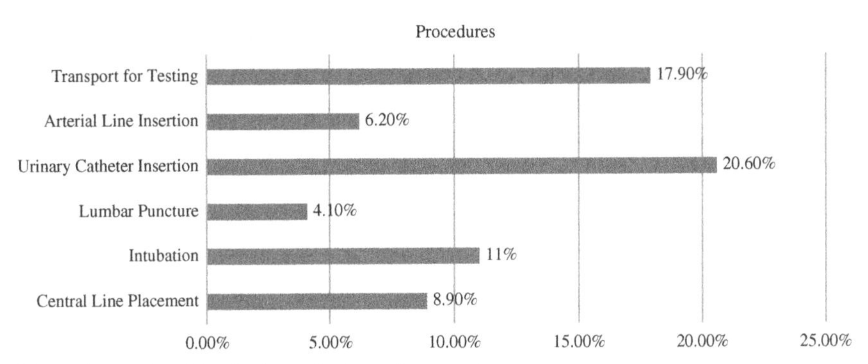

Fig. 7 Example of basic data displays: Bar graph.

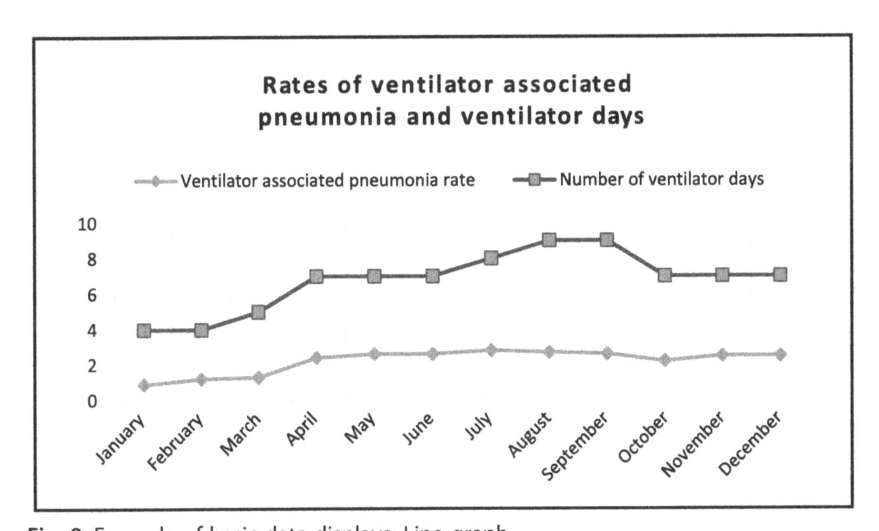

Fig. 8 Example of basic data displays: Line graph.

Data management: QI or research?

There is a common misnomer that QI projects do not rely on statistical analysis because QI is not traditional research. As previously mentioned, it is helpful to think of QI and research on a continuum. The degree to which variables are controlled in the QI project during data analysis helps to inform if the project borrows from research methodology (high control of variables, traditional research design methodologies) or QI (less or no control of potentially confounding variables, may still employ a quasiexperimental design or uncontrolled before-and-after design).[26] Often, the underlying goal of the project drives the design and thus where the project lies on the research and QI continuum. QI projects are generally done with the goal of understanding and/or improving in a local context. Variables that may confound or interact with the phenomena of interest are not controlled, though most QI projects still seek to understand those variables. In traditional research, an effort to predict the impact of an intervention on an outcome mandates the control of variables involved in the phenomena. Both research and QI use some statistical analysis of data to evaluate for change or impact on outcomes.

QI data analysis

Many QI projects rely on descriptive statistics and occasionally, simple inferential statistics to interpret the impact of the project. QI projects that integrate more of a research design may incorporate more advanced statistics such as multivariate analysis. Because QI projects do not rely on power analysis as is common in research studies, many QI leaders struggle with sample size estimates for nonresearch QI work. While there is no absolute rule for sample sizes when doing QI work, small sample sizes can be very revealing and help ensure the "rapid" nature of rapid cycle QI.[27] For example, if a QI leader is looking to identify the compliance with a set of process steps critical in a QI program, a consecutive sample or a systematic random sample (e.g., every third patient admitted who meets a certain criteria) of 20–30 patients may be robust enough to accurately identify gaps in system performance. In fact, small sample sizes may be sufficient in some cases to establish statistical significance in QI outcomes (see Table 6).

One must still pay close attention to selection when using small sample sizes. In order to use small sample sizes with confidence, the QI leader must maintain the integrity of the data sample. Table 6 highlights suggested minimum sample requirements that can be used with QI projects. It is important

Table 6 Estimate sample sizes for QI

Observed system performance (%)	Desired system performance	
	80%	**90%**
95	26	140
90	70	Not applicable
85	260	180
80	Not applicable	50
75	280	28
70	80	20
66	45	15
60	25	10
50	12	6
40	10	5
20	5	5

Minimum sample sizes required for improvement projects based on observed and desired system performance.
The table shows the approximate sample size required to reject the null hypothesis that observed performance (from an audited sample) is consistent with the desired system performance, shown here as being either 80% or 90%. If you wish to calculate an exact *P* value for your audit or Plan-Do-Study-Act (PDSA) result, follow the steps in online supplementary appendix 1. If you wish to calculate the exact 95% CI for your audit or PDSA result, follow the steps in online supplementary appendix 2. The results shown here all use the conventional two-tailed *P* value of .05.
From Etchells E, Ho M, Shojania KG. Value of small sample sizes in rapid-cycle quality improvement projects. *BMJ Qual Saf.* 2016;25(3):202–206, Permission needed.

to remember that when using a small sample size for QI work, several key principles must be considered[27];

- Define the eligibility sample
- Establish exclusion criteria
- State your study period
- Keep a reject log
- Make data collection complete

Advance QI data display

QI projects may benefit from more advanced data analysis and/or display techniques, including statistical control charts, interrupted time series, and Pareto analysis. When used correctly, these tools help to describe phenomena and assist the QI team in responding to variability appropriately. A control chart (also known as statistical control chart or statistical process control) operates from a few key assumptions: processes vary by nature, and that variation may indicate that a process is overall in control or out of control.[28] When normal or expected variation occurs, this is called common-cause variation. Common-cause variation indicates that the process being measured is stable and predictable. This is commonly referred to as being "in control." However, this does not necessarily mean the process is optimized and does not need to be improved. But the scope of the efforts to improve common cause variation would generally be different than when reacting to special cause variation. Special cause variation indicates that the process being measured has changed significantly and is no longer predicable due to some sort of external event. Using control charts helps QI leaders to identify the nature of variation and the type of response warranted.

Control charts are developed using line graphs and simple descriptive statistics. There are several types of control charts, including those that analyze attributes data and those that analyze variables data. However, the u-chart is most commonly used. Control limits are applied to the graph and represent 3 standard deviations in the data above and below the process measured on the line graph. These lines are called upper and lower control limits. At least 20–25 data points must be established for a control chart to be reliable. Finally, control charts generally also identify a trend line. With training, control charts can be developed using simple database software such as Excel or Access. When the process is evaluated in comparison to the upper and lower control limits, any data falling outside the limits are considered special cause variation and indication that the process is "out of control," or some external factor is impacting the process. Fig. 9 displays an example of a control chart that displays the rate of falls for a specific unit over time with control limits:

Interrupted time series (ITS) charts are closely related to statistical process control charts.[28] However, ITS incorporates the before-and-after nature of QI, allowing for an intervention to be noted as the interrupter of the time series. ITS does require more advanced knowledge of statistics as it uses regression modeling. When using ITS modeling, future data points are estimated off of baseline data assuming no intervention takes place, and then

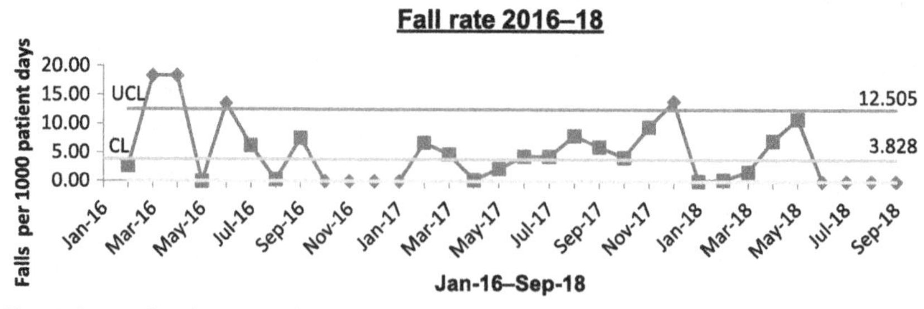

Fig. 9 Example of a control chart with confidence limits.

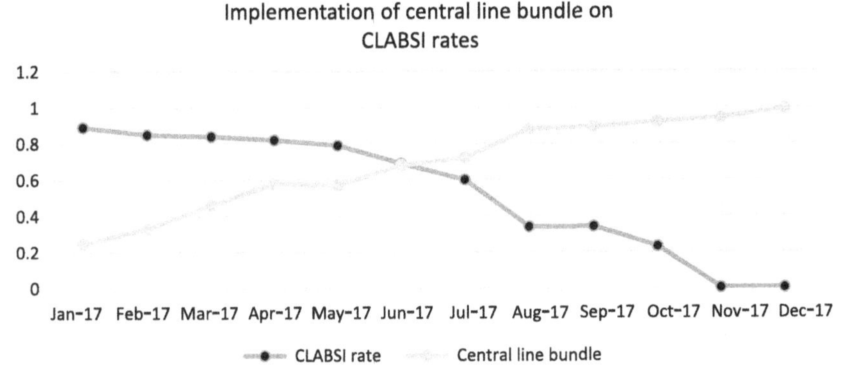

Fig. 10 Example of an interrupted time series chart.

compared against postintervention data to identify if any difference in data is noted. The ITS graph will estimate the level of change and slope of change. While ITS modeling may require more advanced statistical knowledge and application, it is an important tool for QI projects. An example of an ITS series chart that displays the trend in Central Line Associated Blood Stream Infections (CLABSI) rates over time as a CLABSI prevention bundle was implemented is displayed in Fig. 10:

Finally, Pareto analysis is another QI data measurement and display tool that is useful in QI projects. The Pareto principle estimates that 20% of factors account for 80% of success or struggles, highlighting the difference between effort and results. Therefore, it is important to understand within any QI program what 20% of factors are driving the majority of outcomes, be they positive or negative. A Pareto analysis offers a fairly simple approach to evaluate a program. Once a problem area has been identified, factors contributing to the problem are categorized, counted, and then graphed. This activity allows objective analysis of factors contributing to a problem and minimizes gut feelings or hunches from driving improvement activity. Conducting a Pareto analysis can be an important tool and visual representation of the problem to guide QI interventions.

Regardless of the QI methodology used, it is important to employ a systematic approach to the project, including appropriate utilization of data, statistics, and methods for displaying project outcomes. It is equally important to consider how the QI project fits within the larger scope of the initiative under investigation. Specific considerations include: Was the QI project initiated to meet an established benchmark or organizational goal? Was it initiated to incorporate a best practice recommendation? Were the strategies implemented successful at meeting intended targets? Are these strategies sustainable over time and producing reliable and consistent results? Depending on the answers to these questions, ongoing QI projects may be needed to continually evaluate for program effectiveness and assess for necessary modifications over time.

Lastly, it is important to recognize the distinction between QI and EBP-QI. Particularly when QI is initiated under the umbrella of translational science, its purpose is to identify optimal strategies to integrate EBP into routine clinical practices. In this scenario, EBP-QI is utilized, as research has generated evidence to guide practices (EBP), implementation science investigates what impacts uptake of this EBP into routine practice, and QI methodologies can be used to promote consistency of the EBP utilization, identify inefficiencies or factors that impede uptake, and evaluate impact of the EBP on outcomes. The process of QI can also occur outside the realm of translational science and EBP, as QI methodologies can be used throughout health systems to identify root causes of adverse events, system inefficiencies, and to streamline processes. Regardless of whether EBP-QI or QI is used, it is equally important for nurses to understand common methodologies, data generated, and impact on care processes.

Conclusion

As evidenced throughout this book, nurses must be knowledgeable about the different types of data that drive healthcare delivery, such as benchmarks that drive financial reimbursement, indices that reflect quality of nursing care, research that identifies best practices, and IS and QI strategies that promote use of those best practices into routine clinical care. Nurses remain at the forefront of care delivery and represent the largest component of the healthcare workforce. Therefore, knowledge of data and processes for evaluation is critical for nurses to continue to serve as advocates for patients and care processes.

References

1. Joshi MS, Ransom ER, Nash DB, Ransom SB. *The Healthcare Quality Book: Vision, Strategy, and Tools*. 3rd ed. Health Administration Press; 2014.
2. American Association of Colleges of Nursing. The Essentials of Doctoral Education for Advanced Nursing Practice. AACN Reference for DNP program. Retrieved online from: https://www.aacnnursing.org/DNP/DNP-Essentials; 2006.
3. Institute of Medicine. *Crossing the Quality Chasm: A New Health System for the 21st Century*.
4. Institute of Medicine. *To Err Is Human: Building a Safer Health System*. .

5. Institute for Healthcare Improvement. *Free From Harm: Accelerating Patient Safety Improvement Fifteen Years After to Err Is Human.* .
6. Berwick DM, Nolan TW, Whittington J. The triple aim: care, health, and cost. *Health Aff (Project Hope).* 2008;27(3):759–769.
7. Centers for Medicare and Medicaid Services. Core Measure. Retrieved from: https://www.cms.gov/Medicare/Quality; 2018.
8. National Centre for Advancing Translational Sciences, National Institute of Health. Translational Science Spectrum. Retrived online from: https://ncats.nih.gov/translation/spectrum; 2018.
9. Bauer MS, Damschroder L, Hagedorn H, Smith J, Kilbourne AM. An introduction to implementation science for the non-specialist. *BMC Psychol.* 2015;3:32–44.
10. Eccles MP, Mittman BS. Welcome to implementation science. *Implement Sci.* 2006;299: 11–13.
11. Rabin BA, Brownson RC. Developing the terminology for dissemination and implementation research. In: Brownson RC, Colditz GA, Proctor EK, eds. *Dissemination and Implementation Research in Health.* New York: Oxford University Press; 2012.
12. Nilsen P. Making sense of implementation theories, models, and frameworks. *Implement Sci.* 2015;10:53–66.
13. Rogers EM. *Diffusion of Innovations.* 5th ed. New York: Free Press; 2003.
14. Straus S, Graham ID, Zwarenstein M, Bhattacharyya O. Monitoring and evaluating knowledge. In: Tetroe J, Straus S, Tetroe J, Graham ID, eds. *Knowledge Translation in Health Care.* Oxford: Wiley-Blackwell; 2009:151–159.
15. Damschroder LJ, Aron DC, Keith RE, Kirsh SR, Alexander JA, Loery JC. Fostering implementation of health services research findings into practice: a consolidated framework for advancing implementation science. *Implement Sci.* 2009;4:50.
16. Powell BJ, Waltz TJ, Chinman MJ, et al. A refined compilation of implementation strategies: results from the Expert Recommendations for Implementing Change (ERIC) project. *Implement Sci.* 2015;10:21.
17. Michie S, Johnston M, Abraham C, Lwton R, Parker D, Walker A. Making psychological theory useful for implementing evidence based practice: a consensus approach. *Qual Saf Health Care.* 2005;14(1):26–33.
18. Powers WJ, Rabinstein A, Ackerson T, et al. Guidelines for the early management of patients with acute ischemic stroke: a guideline for healthcare professionals form the American Heart Association/American Stroke Association. *Stroke.* 2018;49(3): e49–e99.
19. Padian NS, Holmes CB, McCoy SI, Lyerla R, Bouey PD, Goosby EP. Implementation science for the US President's Emergency Plan for AIDS Relief (PEPFAR). *J Acquir Immune Defic Syndr.* 2011;56(3):199–203.
20. Marshall M, Pronovost P, Dixon-Woods M. Promotion of improvement as a science. *Lancet.* 2013;3831:419–421.
21. United States Department of Health and Human Services, Health Resources and Services Administration. Quality Improvement. Retrieved from: https://www.hrsa.gov/sites/default/files/quality/toolbox/508pdfs/qualityimprovement.pdf; 2011.
22. Donabedian A. Evaluating the quality of medical care. *Milbank Q.* 2005;83:691–729.
23. Glasgow JM, Scott-Caziewell JR, Kaboli PJ. Guiding inpatient quality improvement: a systematic review of Lean and Six Sigma. *Jt Comm J Qual Patient Saf.* 2010;36 (12):533–540.
24. Scoville R, Little K. *Comparing Lean and Quality Improvement.* Institute for Healthcare Improvement White Paper. Cambridge, MA: Institute for Healthcare Improvement; 2014.
25. McQuillan RF, Silver SA, Harel Z, et al. How to measure and interpret quality improvement data. *Clin J Am Soc Nephrol.* 2016;11:908–914.

26. Portela MC, Pronovost PJ, Woodcock T, Carter P, Dixon-Woods M. How to study improvement interventions: a brief overview of possible study types. *Postgrad Med J.* 2015;91(1076):343–354.
27. Etchells E, Ho M, Shojania KG. Value of small sample sizes in rapid-cycle quality improvement projects. *BMJ Qual Saf.* 2016;25(3):202–206.
28. Fretheim A, Tomic O. Statistical process control and interrupted time series: a golden opportunity for impact evaluation in quality improvement. *BMJ Qual Saf.* 2015;24(12): 748–752.

CHAPTER SIX

Protecting the data: Security and privacy

India Eaton*, Molly McNett†
*Institutional Review Board, The MetroHealth System, Cleveland, OH, United States
†The Helene Fuld Health Trust National Institute for Evidence-Based Practice in Nursing & Healthcare, College of Nursing, The Ohio State University, Columbus, OH, United States

Contents

Introduction

Hearing the word "data" may cause individuals to think about a wide variety of topics, from how much data they have used on their phone plans this month to the data that their social media accounts store. Data are defined as "individual facts, statistics, or items of information."[1] In healthcare, the amount of data is increasing, and data that once were only recorded on paper have now shifted to digital records.[2] With this increase in data comes an increased responsibility for the need to secure them safely. This chapter focuses on data security, which includes information about specific privacy and security rules of healthcare data in the United States. These include the Health Insurance Portability and Accountability Act (HIPAA) Privacy Rule and its requirements for protecting data, the HIPAA Security Rule along with the best practices for securing data, and the HIPAA Breach Notification Rule along with the steps to take if a data breach occurs. Additional considerations for data security on an international level will also be presented.

Data for Nurses
https://doi.org/10.1016/B978-0-12-816543-0.00006-6

HIPAA Privacy Rule

The Standards for Privacy of Individually Identifiable Health Information, which is more commonly known as the HIPAA Privacy Rule, were created by the Department of Health and Human Services (DHHS) to ensure that an individual's health data are protected while they are shared within and between hospitals and other covered entities with the purpose of providing healthcare to the individual. "Covered entities" to which the HIPAA Privacy Rule applies include (1) healthcare providers, (2) health plans, and (3) healthcare clearinghouses. Business associates of covered entities whose services involve the use of protected health information (PHI) are also covered under the HIPAA Privacy Rule.[3]

The HIPAA Privacy Rule defines PHI as "'individually identifiable health information' held or transmitted by a covered entity or its business associate, in any form or medium, whether electronic, on paper, or oral."[4] An individual's personal information is only considered PHI if it is specific to or used for healthcare purposes.[4] For example, a person's full name being present on a social media account is not PHI because it is not related to their health; however, a person's name listed with their health condition in a medical chart would be considered PHI. DHHS has identified 18 specific identifiers as being PHI. These 18 identifiers are listed in Table 1.

As a measure of protecting PHI, the HIPAA Privacy Rule has implemented a minimum necessary requirement for covered entities sharing and using PHI. The principle of this requirement is to only obtain the limited amount of PHI needed for the reason it has been requested. This helps to ensure that PHI that is not required is not accessed.[3] The minimum necessary requirement demands that covered entities create and enforce policies that restrict inappropriate access to and use of PHI. For routine disclosures of PHI, covered entities may create policies that limit disclosure of unneeded PHI while still achieving the purpose of the request. However, for non-routine disclosures of PHI (e.g., research purposes), covered entities must review each request on a case-by-case basis, ensuring the request meets their minimum necessary policy and that only the minimum amount of PHI is disclosed for the intended purpose.

For example, a hospital may create a policy that allows for healthcare staff to have unrestricted access to the medical record as a part of their job duties in treating patients; however, the same hospital policy may restrict employees who do not work in patient care from having access to PHI.[6]

Table 1 Specific identifiers of PHI.[5]

- Names
- Any geographic subdivisions smaller than a state (e.g., address, city, zip code, etc.)
- Any elements of dates (not including the year) related to an individual, including ages 90 and older
- Phone numbers
- Vehicle identifiers and serial numbers (e.g., license plate numbers)
- Fax numbers
- Certificate/license numbers
- Device identifiers and serial numbers
- Email addresses
- Web universal resource locators (URLs)
- Social security numbers
- Internet protocol (IP) addresses
- Medical record numbers
- Biometric identifiers (e.g., voice and finger prints)
- Health plan beneficiary numbers
- Full face photographs and comparable images
- Account numbers
- Any other unique identifying information

Similarly, those same healthcare workers who have unrestricted access to the medical record may only be permitted to access the medical record of the patients they are currently treating. Accessing the medical record of patients who are not under their direct care may then be a violation of that policy. This would include accessing their own medical record, or that of family members, or other hospitalized patients not under the healthcare staff's direct care. There are a few circumstances to which the minimum necessary requirement does not apply: (1) when the individual who is the subject of the PHI authorizes the disclosure, (2) when the individual who is the subject of the PHI requests the disclosure for himself or herself, (3) when a healthcare provider uses it for treatment, (4) when the law requires the use or disclosure, (5) when DHHS requests the disclosure to determine compliance with the HIPAA Privacy Rule, and (6) when requested to determine compliance with other HIPAA rules.[7]

HIPAA Security Rule

While the HIPAA Privacy Rule protects the privacy of PHI, the HIPAA Security Rule goes a step further by specifying that physical, technical, and administrative safeguards are needed to secure electronic PHI (e-PHI). The Security Standards for the Protection of Electronic Protected Health Information (i.e., the HIPAA Security Rule) were created in the United States to protect individuals' PHI while permitting covered entities to use technology with the hope of improving patient care. Covered entities are responsible for complying with the HIPAA Security Rule in addition to the HIPAA Privacy Rule. However, the HIPAA Security Rule only applies to e-PHI; it does not cover the other PHI types mentioned in the HIPAA Privacy Rule (i.e., oral or written PHI). The HIPAA Security Rule requires covered entities to (1) ensure the availability, confidentiality, and integrity of all e-PHI they create, keep, receive, and disclose; (2) identify and guard against potential threats against integrity and security; (3) guard against potential nonpermitted uses or disclosures; and (4) ensure their workforce is compliant with these standards.[8]

Methods to secure e-PHI must include physical, technical, and administrative safeguards. Examples of each type of safeguard are listed in Table 2. Physical safeguards to secure e-PHI include facility control and access, which ensures that only limited, authorized individuals are given access to facilities. Device and workstation security not only limits authorized access to electronic media, but also requires implementation of policies regarding removal, transmission, use, and disposal of e-PHI. Technical safeguards to secure e-PHI include access control, which allows only authorized individuals to access e-PHI. Audit controls must be in place to inspect access and other activities that occur within systems using e-PHI; similarly, integrity controls must ensure that e-PHI has not been inappropriately changed or destroyed. Transmission security ensures protection of e-PHI from unauthorized access while it is being electronically transmitted over a

Table 2 Methods to secure e-PHI.

Physical safeguards	Technical safeguards	Administrative safeguards
• Facility control and access • Device and workstation security	• Access control • Audit controls • Integrity controls • Transmission security	• Security management processes • Security personnel • Information access management • Workforce training and management • Evaluation processes

network. Lastly, administrative safeguards to secure e-PHI include a security management process, in which potential risks to e-PHI are identified and measures are taken to reduce these risks. Security personnel must also be present to create, implement, and oversee the security policies. Information access management processes must ensure that access to e-PHI is only authorized when appropriate for the user's or recipient's role. Workforce training and management mandates all covered entities train all employees regarding the security policies and procedures as well as supervise employees who have access to e-PHI as a part of their job duties. Finally, an evaluation component must ensure a covered entity periodically reviews its security policies and determines compliance with current HIPAA Security Rules.[8]

Securing research and quality improvement data

In the United States, healthcare research data that include any PHI must follow the protections outlined earlier. In addition to these protections, the National Institutes of Health (NIH) has a policy in place for NIH grantees regarding how to secure research data; however, many regulatory agencies apply the principles of this policy to data security for all studies, and not just those funded by the NIH. For research data, the NIH recommends that identifiable data not be stored on portable devices (e.g., laptops, flash drives, etc.); however, if this storage method must be chosen, then the data should be encrypted.[9] To encrypt data means to code the data in such a way that only authorized individuals can read it.[10] The NIH also recommends that password protection be used to limit access to identifiable information. When data must be transferred outside of an institution, it should only be done when it is known that the recipient institution's systems are secure.[9]

The protections for PHI data apply to data collected for research, quality improvement, and reporting purposes. As such, there are many practical ways to secure PHI and other research or quality improvement data. Whenever possible, data should be deidentified, with any identifying link between the data and PHI stored in a separate location from the data. Fig. 1 provides an example of how to deidentify data using a study or project log. Each study record is assigned a sequential number (i.e., 1, 2, 3, etc.) and paired with the identifier, which is typically a patient medical record number. This log book is stored securely according to the regulations and is then kept separate from the rest of the data that contains information about the patient or subject's health condition (i.e., age, gender, diagnosis, complications, etc.). The resulting dataset then only contains the sequential number, often called

STUDYID	MEDRECORD
1	555666
2	234567
3	123456
4	456789
5	234567
6	789023
7	677890
8	654345
9	654123
10	123678

Fig. 1 Example of study log.

record_id	age	gndr	ethnicity
1	36	0	2.00
2	61	1	1.00
3	73	0	1.00
4	69	0	1.00
5	55	0	2.00
7	75	0	2.00
8	53	1	1.00
9	35	1	1.00
10	66	0	1.00

Fig. 2 Example of deidentified study database.

the "study identifier" or "record identifier," and the relevant data for the project, without any PHI (Fig. 2).

Specifically, for quality improvement data and benchmark reporting, aggregate reports of outcome data should only be captured without specific patient identifiers included (Fig. 3). In research and quality improvement data activities, only study or project staff members should have access to identifiable information. If data must be sent outside of an institution, a confidentiality agreement should be signed between the institution and the recipient of the data, and methods for encryption of the data being sent should be employed.[11] Access reviews should be performed on occasion to ensure that only individuals currently working with the data have access

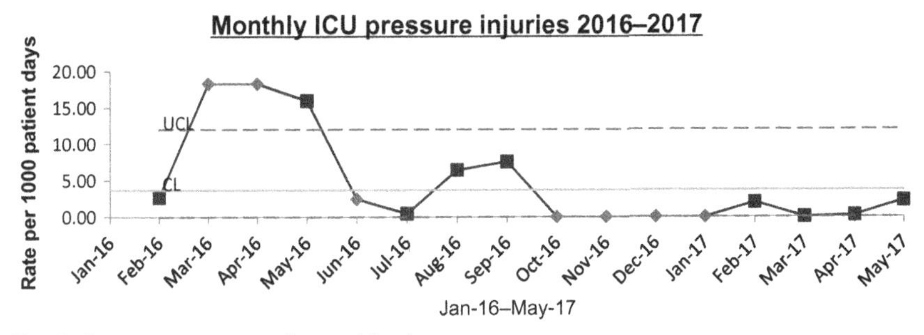

Fig. 3 Aggregate report of monthly data.

to it.[12] Any changes made to improve data security should be documented in accordance with HIPAA.[13] Paper records should be "double-locked" whenever possible, meaning that they are placed in a locked cabinet in a locked room when not in use. If double-locking is not possible, then paper records should, at the very minimum, be locked in a secure location away from public access.[14] While it is recommended to destroy identifiable information as soon as possible, this may not always be appropriate. For example, in research, DHHS requires that data be retained for a minimum of 3 years after completion of a study; however, some institutions may require longer retention periods. Once the data retention period has ended, assurance must be made that data are thoroughly deleted and destroyed. Data should not be able to be reobtained or recreated.[15]

Data breaches

In a perfect world, securing PHI and other data would happen without error; however, data breaches do occur. Between 2008 and 2014, a total of 8570 incidents related to human subjects research were reported to the Office of Human Research Protections (OHRP), which is the office within DHHS in charge of ensuring institutions remain compliant with the human subjects research regulations. Of these incidents, 582 were deemed to be risks to or breaches of confidentiality.[16] Breach of confidentiality is defined by DHHS as "an impermissible use or disclosure under the Privacy Rule that compromises the security or privacy of the protected health information."[17] Breaches of confidentiality should not be confused with incidental uses or disclosures of PHI, which occur as a consequence of sharing PHI for a required, permitted reason that complies with the HIPAA Privacy Rule. Incidental uses or disclosures of PHI are unable to practically be prevented, are limited in nature, and do not violate the HIPAA Privacy Rule. For example, a patient's visitor at a hospital overhearing the patient's provider discuss confidential information with the patient would not be considered a breach of confidentiality, but rather an incidental disclosure.[18]

When a breach occurs, covered entities in the United States are required to comply with the HIPAA Breach Notification Rule, which states that affected individuals, the DHHS Secretary, and the media (when applicable) must be notified of the breach. Breach notifications must include a description of the breach, the types of information affected by the breach, what affected individuals need to do to protect themselves from harm, the steps that the covered entity is taking to investigate the breach as well as prevent it from occurring again, and the contact information of the covered entity. Affected individuals must be notified in writing (i.e., either by mail or email) within 60 days of the breach. If 10 or fewer affected individuals have outdated contact information, then the covered entity may contact these individuals by an alternative means (e.g., telephone). However, if more than 10 affected individuals have outdated contact information, then the covered entity is required to give alternative notice via their website or through broadcast media where the affected individuals live. A toll-free phone number must be given in these notices so that the individuals can determine if they were affected by the breach. The phone number must be available for a minimum of 90 days. When a breach affects more than 500 individuals, covered entities are required to inform the media and DHHS Secretary within 60 days of the breach.[19] Also, the DHHS Secretary is required to publicly post these breaches for 24 months.[20] However, when fewer than 500 individuals are affected, the media does not need to be notified unless issues arise regarding outdated contact information for affected individuals as described earlier. Also, in this case, the DHHS Secretary would only be notified no later than 60 days after the end of the calendar year when the breach is discovered. Covered entities are responsible for proving that the appropriate notifications have occurred.[21]

Case example

Some data breaches are obvious, such as a computer hacker getting into a hospital's electronic medical record and obtaining patients' PHI. However, some breaches may not be as apparent. For example, a hospital employee is making rounds and notices that one of the newly admitted patients is her friend's cousin. Not sure if her friend knows that the cousin has been admitted to the hospital, the hospital employee calls her friend to inform her of this information. The hospital employee does not stop to obtain permission from the cousin before calling her friend. She also does not check to see that the cousin specifically requested that his information be kept out of the patient directory. Although the hospital employee only wanted to inform her friend, this would be considered a breach of confidentiality. To prevent this from being a breach of confidentiality, prior to disclosing the patient's information, the hospital employee should have obtained authorization from the patient to share this information. The hospital employee also could have checked to confirm that this patient's information had not been requested to be removed from the patient directory. If either of these steps had been taken before the disclosure, then this situation would not have been a breach. However, since neither of these actions was taken, the hospital employee disclosed information about the cousin (i.e., the patient) that was not permitted under the HIPAA Privacy Rule. Specifically, the patient's admittance to the hospital was individual health data that were to be protected within the confines of the hospital. As a result, once the breach was discovered, the hospital would then be responsible for notifying the patient (i.e., the cousin) about the breach in writing within 60 days. The hospital would also be responsible for reporting the breach to the DHHS Secretary no later than 60 days after the end of the calendar year when the breach was discovered. While it may seem more reasonable for the computer hacking breach to be reported as compared to the example regarding the hospital employee, the HIPAA reporting requirements are the same and therefore apply to all breaches.

Case example

A hospital employee was newly appointed to be the quality designee for their department and was tasked with gathering data on all hospital readmissions for their unit. The individual was supplied with a list of patients from the hospital quality department of individuals who were readmitted to the hospital within 30 days of discharge from the employee's unit. The employee was instructed to examine the medical records of those patients to identify the reason for the readmission. In order to track this information, the hospital employee created an excel spreadsheet on their laptop. The first column in the spreadsheet contained the patient's medical record number and original admission date. The employee then created new columns to record the date of readmission, and reason for readmission that they gathered from the medical record of the patient. The employee kept the laptop in their car or at the unit front desk to continue recording information on reasons for readmission.

Questions:
What are potential sources for data breaches in this scenario?
What protections should be in place to secure this data?
What are the PHI elements in this scenario?
What additional considerations or approvals should be obtained to access this PHI?
What are alternative methods for gathering and securing this type of data?
What are possible consequences for the individual using this technique to store the data?
What are possible consequences for the organization and the patient?

International considerations

While the majority of this chapter focuses on data protection regulations within the United States, specific countries may have additional laws in place to add further protections for healthcare and consumer data. In the European Union, for example, the General Data Protection Regulation (GDPR) includes many of the data protection practices inherent in HIPAA, yet adds additional measures applicable to all citizens, not just those receiving healthcare from a covered entity within a certain period of time.[22] Under the regulations, health data are categorized as Data Concerning Health, Genetic Data, and Biometric Data. Data from any of these categories can only be utilized if the individual provides consent; if data are necessary for a medical diagnosis, preventative health, or provision of health services; or if data are required for public health purposes. The GDPR is ultimately structured to allow individuals to have control over how their data are monitored and utilized, as well as for what period of time. Table 3 highlights key differences between GDPR and HIPAA regulations.

While regulations for healthcare data may vary by country, it is ultimately the responsibility of the individual healthcare provider or individual accessing the data to be knowledgeable about any applicable regulations by law, as well as any additional policies and procedures for data utilization within specific institutions. There may be institutional, geographical, and temporal variations of data security practices; yet, the key tenet of ensuring protections for privacy, security, and confidentiality of individual data remains essential across all settings.

Table 3 Differences between HIPAA and GDPR regulations.[22]

	HIPAA	GDPR
Purpose	Protect identifiable information of patients. Create uniform standards for use of patient health information.	Protect all personal data, not just healthcare data. Provide a mechanism for secured utilization of personal data throughout EU. Standardize data protection across the EU.
Scope	Healthcare entities	All industrial entities
Type of data	Identifiable, individual health information.	Personal data, sensitive genetic or biometric personal data.
Breach notification	If breach occurs, must notify within 60 days.	If breach occurs, must notify within 72 hours.

EU, European Union.

Case example

Nurses in an ICU have noted an increase in the number of pressure injuries (PI) among their patients in recent months. A quality improvement project included integration of Plan-Do-Study-Act (PDSA) cycles to address common factors, such as more accurate screening and documentation of skin assessments, and improved turning practices every 2 hours that were suspected to contribute to the PI development. Aggregate reports on the incidence of PI were provided to the nurses from the hospital quality department every month to evaluate if their interventions were successful. These reports did not include any PHI; rather, they just included monthly numbers of patients with PI.

However, monthly reports indicated the rates of PI continued to increase. Therefore, the nurses consulted with an on-site research mentor and designed a research study to evaluate prevalence and association of specific risk factors with PI development. The group received IRB approval for their study. They also received HIPAA authorization to access the patients' medical records and gather PHI as part of their study. The nurses needed medical record numbers of patients to look up causative factors for PI. In order to deidentify their dataset, they created a study log book to keep the medical records (i.e., PHI) separate from their database. Each subject was assigned a consecutive study identification number, such as 1, 2, and 3. This identification number was then logged to their medical record number (Fig. 1).

As a result, the subjects in the database were identified by the study ID, or record ID number, and not by their medical record number, as an added measure to ensure data security (Fig. 2).

The log was stored in a separate location from the database. Therefore, if the database was breached, there was no identifying information present. Similarly, if the log was breached, the medical record number would be present, but it would not be linked to specific health information of that patient. Both electronic versions of these logs were stored on password protected, encrypted, secure network drives within the health system computer servers to further decrease risk of a data breach.

Conclusion

For many, access to PHI and other data is a necessary function of employment and other related activities that cannot be avoided. Securing PHI and other data may appear to be an intimidating task, especially when considering the vast amount of data available for access. One may be hesitant about the potential for data breaches as well as the consequences associated with them. However, knowing the regulations and applying the best practices for data security can make this a less daunting experience.

References

1. Data. *Dictionary.com*. https://www.dictionary.com/browse/data. Accessed 30 December 2018.
2. Patil HK, Seshadri R. Big data security and privacy issues in healthcare. 2014 IEEE International Congress on Big Data 2014. Doi: https://doi.org/10.1109/BigData.Congress.2014.112. [Accessed December 30, 2018].
3. Office for Civil Rights (OCR). *Summary of the HIPAA Privacy Rule. HHS.gov*. https://www.hhs.gov/hipaa/for-professionals/privacy/laws-regulations/index.html. Accessed 19 December 2018.
4. Office for Civil Rights (OCR). *Guidance Regarding Methods for De-Identification of Protected Health Information in Accordance with the Health Insurance Portability and Accountability Act (HIPAA) Privacy Rule. HHS.gov*. https://www.hhs.gov/hipaa/for-professionals/privacy/special-topics/de-identification/index.html. Accessed 19 December 2018.
5. U.S. Department of Veterans Affairs. *List of 18 HIPAA Identifiers. VA.gov*. https://www.atlanta.va.gov/Docs/HIPAA_Identifiers.pdf. Accessed 23 December 2018.
6. Office for Civil Rights (OCR). *Minimum Necessary Requirement. HHS.gov*. https://www.hhs.gov/hipaa/for-professionals/privacy/guidance/minimum-necessary-requirement/index.html. Accessed 19 December 2018.
7. Department of Health and Human Services. *Protecting Personal Health Information in Research: Understanding the HIPAA Privacy Rule. NIH.gov*. https://privacyruleandresearch.nih.gov/pdf/HIPAA_Privacy_Rule_Booklet.pdf. Accessed 20 December 2018.
8. Office for Civil Rights (OCR). *Summary of the HIPAA Security Rule. HHS.gov*. https://www.hhs.gov/hipaa/for-professionals/security/laws-regulations/index.html. Accessed 26 December 2018.

9. National Institutes of Health. *NIH Grants Policy Statement. NIH.gov.* https://grants.nih.gov/grants/policy/nihgps_2013/nihgps_ch2.htm#protecting_sensitive_data. Accessed 29 December 2018.
10. University of Pittsburgh. *Electronic Data Security. Pitt.edu.* https://www.irb.pitt.edu/electronic-data-security. Accessed 29 December 2018.
11. University of Michigan. *Data Security Guidelines. Umich.edu.* https://research-compliance.umich.edu/data-security-guidelines. Accessed 29 December 2018.
12. University of Massachusetts Medical School. *Securing Research Data. UMassMed.edu.* https://www.umassmed.edu/it/policies-and-guidelines/best-practices/securing-research-data/. Accessed 29 December 2018.
13. i.e., in reference to 164.310(a)(2)(iv). Department of Health and Human Services. *Security Standards: Physical Safeguards. HHS.gov.* https://www.hhs.gov/sites/default/files/ocr/privacy/hipaa/administrative/securityrule/physsafeguards.pdf?language=es. Accessed 29 December 2018.
14. Clinical Tools, Inc. *Guidelines for Responsible Data Management in Scientific Research. HHS.gov.* https://ori.hhs.gov/images/ddblock/data.pdf. Accessed 29 December 2018.
15. i.e., in reference to 45CFR46.115(9)(b). Clinical Tools, Inc. *Guidelines for Responsible Data Management in Scientific Research. HHS.gov.* https://ori.hhs.gov/images/ddblock/data.pdf. Accessed 29 December 2018.
16. Ramnath K, Cheaves S, Buchanan L, Borror K, Banks-Shields M. *Incident reports and corrective actions received by OHRP. IRB Ethics Hum Res.* 2016;38:6. https://www.thehastingscenter.org/irb_article/incident-reports-corrective-actions-received-ohrp/. Accessed 12 December 2018.
17. Office for Civil Rights (OCR). *Breach Notification Rule. HHS.gov.* https://www.hhs.gov/hipaa/for-professionals/breach-notification/index.html. Accessed 29 December 2018.
18. i.e., in reference to 45 CFR 164.502(a)(1)(iii). Office for Civil Rights (OCR). *Incidental Uses and Disclosures. HHS.gov.* https://www.hhs.gov/hipaa/for-professionals/privacy/guidance/incidental-uses-and-disclosures/index.html. Accessed 30 December 2018.
19. i.e., in reference to 45 CFR §§ 164.400-414. Office for Civil Rights (OCR). *Breach Notification Rule. HHS.gov.* https://www.hhs.gov/hipaa/for-professionals/breach-notification/index.html. Accessed 30 December 2018.
20. i.e., in reference to section 13402(e)(4) of the HITECH Act. Office for Civil Rights (OCR). *Cases Currently Under Investigation. HHS.gov.* https://ocrportal.hhs.gov/ocr/breach/breach_report.jsf. Accessed 30 December 2018.
21. Office for Civil Rights (OCR). *Breach Notification Rule. HHS.gov.* https://www.hhs.gov/hipaa/for-professionals/breach-notification/index.html. Accessed 30 December 2018.
22. The European Union. *Data Protection Act.* Available from: http://www.legislation.gov.uk/ukpga/2018/12/pdfs/ukpga_20180012_en.pdf; 2018.

CHAPTER SEVEN

Determining the dissemination plan: Internal and external considerations

Wendy Sarver*, Molly McNett†
*Department of Nursing Research, MetroHealth System, Cleveland, OH, United States
†The Helene Fuld Health Trust National Institute for Evidence-Based Practice in Nursing & Healthcare, College of Nursing, The Ohio State University, Columbus, OH, United States

Contents

Data dissemination

Dissemination refers to making knowledge about your project available and accessible to target audiences[1]; this may also be called "contributing to generalizable knowledge," or to what is currently known about a topic. Dissemination is a crucial step in translational knowledge and data utilization. If findings from research and some quality improvement activities or benchmarking efforts are not made known to target audiences or key stakeholders, then knowledge about effectiveness of interventions, risk factors, or methods to optimize outcomes is stifled.[2] Simply put, dissemination is making project findings known to the people who require this information to advance care delivery, replicate studies to strengthen the evidence base, or to implement changes in practice.[3]

Evidence-based practice is integration of the best-available evidence to guide decisions about care delivery and incorporates most relevant evidence, clinical expertise, and patient preferences and values.[4, 5] Data disseminated

Data for Nurses
https://doi.org/10.1016/B978-0-12-816543-0.00007-8

from research and some quality improvement initiatives provide the "evidence" in evidence-based practice. It is the core component to providing care based on information about best practices.

Historically, there have been significant lags in time for integration of evidence into practice, including up to 15–20-year gaps for evidence to reach clinical practice implementation.[2, 4] Efforts in recent years have significantly decreased the 20-year gap between dissemination and practice integration. These efforts include a focus on implementation and dissemination science, creation of national repositories of clinical practice guidelines and systematic reviews, and standardizing methodologies for synthesis of findings across studies into evidence-based recommendations. These initiatives have made data readily accessible and more meaningful for practice, yet the need for ongoing dissemination of data remains critical to continue advancing data utilization into practice. Dissemination fuels development of new clinical practices and serves as validation for current practices across patient populations.[3]

Establishing the dissemination plan

The final phase of any project includes creating a dissemination plan to share key findings, conclusions, or recommendations. Regardless of the type of project, it is important that key stakeholders are informed of relevant results. These results may be used in decisions about direct patient care, updates to policies or procedures, workflow processes, and even financial or strategic planning decisions. Identifying the target audience for your dissemination is a critical first step in the dissemination plan.

Target audiences for dissemination may include individuals within your organization, or groups outside of your organization. Therefore, most dissemination plans include both internal and external dissemination. Internal dissemination is sharing project results within your organization, while external dissemination includes sharing information to individuals or groups outside of the organization in which you work. For both types of dissemination, it is important to ensure any necessary approvals have been obtained to share the data you have to target audiences.

Oftentimes projects may require approval prior to collection of any data. Many organizations have Institutional Review Boards (IRBs) in place that review any project that meets federal definitions of research, prior to project commencement. Generally, if the project has received IRB review, results may be disseminated both internally and externally. Many external venues for dissemination, such as journals and conference abstracts, require documentation of IRB approval prior to accepting content for dissemination. For projects that did not require IRB review, some organizations have other approval mechanisms in place to ensure any data that are disseminated meet privacy requirements and don't adversely impact the organization if data are shared. These approval mechanisms may consist of approval from a department chairperson, direct supervisor, a shared governance council, or a research and/or evidence-based practice council. Once the appropriate approvals have been obtained, discuss the dissemination plan with the proper management or administrative personnel. This is necessary to verify key stakeholder groups, to secure financial resources for budgetary needs if conference attendance is expected, and to ensure dissemination plans align with any other organizational requirements.

Internal dissemination

Once the target audience and appropriate approvals are obtained, the internal dissemination process begins. Internal dissemination may consist of presenting findings at interdisciplinary stakeholder meetings, grand rounds, journal clubs, or internal conference venues. Internal dissemination is important as key stakeholders may make decisions to change policy, practice, or priority areas based on strength and applicability of the project findings. Internal dissemination also facilitates replication and validation of findings across care settings. For example, if an intervention was implemented in one patient care area and resulted in substantially improved positive outcomes, other care areas may be experiencing similar scenarios, and may want to replicate this intervention to determine if it is as effective among their patient populations. Internal dissemination should always be considered an important step in sharing your findings, as this may be the catalyst for change within the organization.

Case example and discussion item

Jane and Steven are nurses working on an adult medical surgical unit. They recently completed a quality improvement project that was approved by their shared governance council. The project goal was to reduce falls on their unit by 50% over a 6-month period by integrating a fall screening tool, fall prevention interventions, and immediate root cause analysis on all falls within 24 hours. Their project was successful, with an overall reduction in falls of 60%. Steven and Jane would now like to disseminate their data. What are some key stakeholder groups that would benefit from internal dissemination of this information?

What would their next steps be?

Table 1 Tips to consider for internal dissemination at stakeholder meetings.

- Who typically attends the meeting
- Amount of time allotted to share findings
- How many participants typically attend
- What is the audience's level of knowledge on the topic?
- What is important to the stakeholders, and what information do they need to make a decision regarding a practice change or allocation of resources?
- Is there access to a computer and/or audio-visual equipment?
- Are handouts recommended?
- Is there a mechanism to share information with participants who may not be present at the meeting?
- Frame your presentation content to the audience and their knowledge level when preparing the presentation.
- Include key points that center on the background/rationale for the project, specific project goals, methods, and results that would be important to your target audience.
- Include conclusions or recommendations that would be of specific importance to this group.

Examples of stakeholder meetings in which to internally disseminate project findings may include a shared governance meeting, nurse management or leadership meetings, or unit-based councils, interdisciplinary rounds, or journal club meetings. Key stakeholders may include interdisciplinary directors, managers, nurses, or other pertinent healthcare providers. Table 1 lists important items to consider when preparing for dissemination to key stakeholders internally.

External dissemination

External dissemination refers to sharing findings from your project with individuals or groups outside of your organization. This can include local, regional, national, and international audiences. External dissemination is critical for translational knowledge development, implementation science, and ultimately establishing the evidence base required to advance care delivery and improve outcomes for patients. Evidence-based practice cannot exist without external dissemination of evidence. Avenues for external dissemination can include poster or oral presentations of findings at conferences or among target audiences outside of your organization, publication in peer-reviewed journals, and in some cases, media reports.

Case example

After presenting their information at key stakeholder meetings within their organization, Jane and Steven are encouraged to disseminate their findings externally by submitting an abstract for a poster presentation at a national conference, and as a manuscript for publication in a peer-reviewed journal. They aren't sure where to start, as they have never done this before. What are some important considerations as they prepare material for dissemination outside of their organization?

What would their next steps be?

Poster and oral presentations

External dissemination of findings via a poster or oral presentation is a vital component of any dissemination plan. When preparing for this mechanism of external dissemination, there are several important considerations. Table 2 lists key questions that should be addressed to aid in identifying the optimal venue for dissemination.

Most external dissemination venues require submission of an abstract or summary of your project prior to accepting it for a poster or oral presentation. Oftentimes there are specific fields required in the abstract, and these may vary depending on the venue. Generally, most abstracts will include: (1) a background section that describes the scope of the problem or issue that led to your investigation; (2) identification of project aims or goals; (3) the methods or intervention performed to address the issue; (4) project findings or results; (5) the implications of these findings for the target audience and/or patient care. There may be additional instructions for abstract submissions, such as a word or character limit, learning objectives, and/or key references. Pay attention to all requirements, as abstracts may be rejected solely for failure to adhere to submission instructions.

Submission instructions include deadlines for submission, dates for notification of acceptance, and author requirements. Notification of acceptance or denial of the submission will also typically include additional instructions. If the abstract is accepted for a presentation, you may have to confirm your acceptance and commit to attending the venue to disseminate your results. Often this confirmation must occur within several days, so be sure to note specific instructions. Instructions also include specifications on poster size (if applicable), utilization of a specific template for oral presentations, required

Table 2 Key questions when preparing for external dissemination via poster or oral presentation.

- ➢ Who is your target audience?
- ➢ Is there a mechanism to disseminate findings to this group (i.e., a local, regional, national, or international meeting or conference)?
- ➢ If there is a mechanism for dissemination, what are the requirements?
- ➢ Is an abstract submission required?
- ➢ What are the abstract acceptance dates?
- ➢ What information is required in the abstract?
- ➢ When are the dates, and are you able to attend?
- ➢ Will your organization allow you the time off work to attend this venue?
- ➢ Are there costs associated with submitting an abstract, or attending the venue?
- ➢ If there are costs, who is responsible for paying you, your organization?
- ➢ If your organization will cover costs, what is the process for approval?
- ➢ Does the abstract need to be reviewed by your department, a separate department, or supervisor before submission?
- ➢ Is there a standardized template for posters or presentation slides that your organization requires you to use for all presentations?
- ➢ What is the process for creating a poster, and are there fees associated with printing?
- ➢ Is there a mentor who can assist you in abstract preparation, presentation materials, and navigating required approvals?
- ➢ Who typically attends the meeting?
- ➢ Amount of time allotted to share findings.
- ➢ How many participants typically attend?
- ➢ What is the audience's level of knowledge on the topic?
- ➢ Is there access to a computer and/or audio-visual equipment?
- ➢ Are handouts recommended?
- ➢ Is there a mechanism to share information with participants who may not be present at the meeting?
- ➢ Consider the target audience and knowledge level when preparing the presentation.
- ➢ Include key points that center on the background/rationale for the project, specific project goals, methods, and results that would be important to your target audience.
- ➢ Include conclusions or recommendations that would be of specific importance to this group.
- ➢ Anticipate potential questions stakeholder groups may have, and prepare additional data or responses to address these potential questions.

times for presentation of findings, deadlines for venue registration, and any costs associated with attendance. When preparing your materials for external dissemination, engage a mentor if needed to assist with generating presentation materials, including posters, presentation slides, and handouts. As with internal dissemination, it is important to consider the audience that will be present at the venue, the time allotted for your presentation, and most importantly, how your findings contribute to advancing evidence-based practice for that field.

Publication in peer-reviewed journals

Peer-reviewed journals offer nurses the opportunity to publish a rigorously reviewed manuscript, which will reach the appropriate audience for the topic and disseminate evidence to guide practice. When preparing for external dissemination via publication in a scholarly journal, again consider your target audience—who would best benefit from the findings of your project? Consider professional nursing or interdisciplinary organizations and investigate if they have a journal that routinely publishes material for those target groups.

If your project findings are applicable to a specific target audience, such as pediatric nurses, consider a pediatric nursing journal, for example. In contrast, if your findings are more applicable to a broader audience, such as project investigating fall prevention across all settings, consider a journal with a broader audience base that would benefit from this information. Generate a list of potential journals with audience types that would benefit from your project findings. Once you have composed a list of several potential journals, visit each journal's website and review the journal scope, types of manuscripts typically published, and information for authors. It is important to also review the previous issues of that journal to see what types of articles have recently been published, previous writing styles, and formatting of content. Based on the review of these criteria, narrow down your list of potential journals. Some journals recommend sending a query to the editor to determine suitability of material for publication. If this is offered, consider writing the editor to determine if your project is something that aligns with their target audience, journal scope, and other requirements. Table 3 lists considerations when selecting a journal for external dissemination.

Once you have decided on a journal, engage a mentor if you have not previously published material in a scholarly journal. Similar to abstract submissions, most journals will require the material to be presented in a structured format, which typically includes background, aims, methods, results, and implications for practice. Be sure to follow the journal author instructions for headings, references, and submission of materials. Prior to submission, ensure any approvals from your organization are obtained. These may include approval from a specific department, supervisor, or committee. Any

Table 3 Key questions when preparing for external dissemination via publication in peer-reviewed journal.

- Who is the target audience?
- Is the target audience narrow or broad?
- Is the target audience only nurses, or interdisciplinary teams?
- What are possible journals that specifically include these target audiences?
- What is the scope, instructions, and submission requirements for these journals?
- Does the journal accept the type of manuscript that is being considered?
- What are the submission requirements, including page limits, writing style, reference style, and formatting?
- What is the timeline for acceptance?
- Has the journal published other articles similar to this project?
- Are there fees associated with publishing in this journal?
- Is the journal reputable and include a rigorous peer review process?
- Are any approvals required from your organization or supervisors prior to submitting the project for publication?
- Is a mentor available to assist with the submission and publication process?

data that are disseminated externally should have approval from some level within the organization prior to submission for publication to ensure protection of privacy, proprietary information, and confidential information.

Once the necessary approvals for submission of a manuscript for publication are obtained, follow the journal instructions for submission. After submission, it often takes 2–4 months for journal reviewers to provide feedback. Once feedback has been completed and summarized, the editor will share this information with the corresponding (or submitting) author. Most journals will either provide feedback that: (1) the manuscript does not align with journal requirements or target audience and is rejected; (2) the manuscript may be accepted if specific changes are made based on reviewer feedback; (3) the manuscript is accepted as submitted. The majority of manuscripts fall under the second category and require some level of editing before being accepted for publication. Specific recommendations for these edits are provided in the reviewer feedback. It is important to pay specific attention to this feedback and required timelines for resubmission. Often a point-by-point response is required with the edited manuscript to specifically highlight the changes that were made as requested by reviewers.

If the manuscript is accepted for publication, you will receive notification, along with an anticipated publication date. Immediately prior to publication, an author "proof" will be sent, which displays the final formatting and content of the article immediately prior to publication. This proof must be reviewed and any last edits made before returning for final publication. Again, it is important to adhere to associated instructions and deadlines for review of this proof in order to ensure timely publication of content. Once the article is published, be sure to share with colleagues and applicable areas within the organization.

If the manuscript is not accepted for publication, do not be discouraged. Often the content of the project may not align with journal priorities or target audience. If this is the case, consider any feedback received from the original journal that declined the project. If there are specific edits that were suggested, consider this feedback and make revisions if applicable. Then, return to the list of potential journals and consider submission to an alternate journal. Again review the instructions for authors, as these vary among most journals. Edit the content of your manuscript as needed to confirm to submission requirements and target journal audience if applicable. When edits are complete, submit the manuscript following the instructions provided.

Additional considerations for external dissemination

When disseminating results externally, be mindful of results that will be shared with wider audiences. For most IRB-approved projects, dissemination of data is acceptable, provided any necessary departmental approvals are obtained. For quality improvement or evidence-based practice projects, consider if the data reported (1) are publicly reported and already available to the public; (2) are in alignment with best-practice standards; (3) reflect quality care provided by the organization. For publicly reported measures, data on outcomes such as external benchmarks may already be required to be reported publicly. If this is the case, it may be acceptable for reporting of data, particularly if your project demonstrated substantial improvements in rates because of an intervention. However, if the outcome data of the project are not publicly reportable, and reflect rates that are significantly below suggested benchmarks, consider presenting data in aggregate form and reporting percent change in rates, rather than actual rates. Lastly, it is equally important to ensure your project is in alignment with established standards of care and reflects care quality of the organization. For any project, substandard care is not acceptable, and a focus on care quality should always be emphasized. External dissemination should highlight the improvement efforts with the organization, and how specific changes resulted in substantial improvements to contribute to generalizable knowledge and optimize care outcomes.

References

1. Lia-Hoagberg B, Schaffer M, Strohschein S. Public health nursing practice guidelines: an evaluation of dissemination and use. *Public Health Nurs*. 1999;16(6):397–404.
2. World Health Organization (WHO). *Quality and Accreditation in Health Care Services: A Global Review*. Switzerland: WHO; 2003.
3. Oermann MH, Shaw-Kokot JS, Knafl GJ, Dowell J. Dissemination of research into clinical literature. *J Clin Nurs*. 2010;19(23–24):3435–3442. https://doi.org/10.1111/j.1365-2702.2010.03427.x.
4. Westfall J, Mold J, Fagnan L. Practice-based research—"Blue Highways" on the NIH roadmap. *JAMA*. 2007;297:403–406.
5. Melnyk B, Fineout-Overholt E. *Evidence-Based Practice in Nursing & Healthcare: A Guide to Best Practice*. 4th ed. Philadelphia, PA: Lippincott, Williams, & Wilkins; 2018.

International perspectives on data

Molly McNett
The Helene Fuld Health Trust National Institute for Evidence-Based Practice in Nursing & Healthcare, College of Nursing, The Ohio State University, Columbus, OH, United States

Contents

Introduction

On an international level, nurses are not immune to the importance of data to drive care delivery. Efforts to improve population health on a global scale were initiated in the 1970s, with the specific goal to ensure accessibility to healthcare to all persons by the year 2000.[1] Making healthcare accessible to all regions is a surmountable goal. Establishing methods to track data and outcomes related to that care, while ensuring the quality of care provided, is another challenge. Once again, nurses serve at the forefront of care delivery to all regions. As such, they are instrumental in evaluating data on processes and outcomes related to quality of care.

Early international benchmarking

Initial recommendations by the World Health Organization (WHO) to ensure accessibility to healthcare include targets specifically aimed at ensuring quality of care.[1] The European region of the WHO advocates for specific structures and processes to be present within developing health systems to provide data to continually evaluate care, highlight areas for improvement, and evaluate impact of technological integration.[1] While initial targets for quality were established in 1984, they were revised and

Data for Nurses
https://doi.org/10.1016/B978-0-12-816543-0.00008-X

reinforced throughout the next decade as healthcare accessibility and evolution of methods for data acquisition and reporting expanded.

In the early 2000s, it became apparent that delineating specific targets for quality of care on a global scale could be challenging, as not all areas are equipped with the similar resources. A seminal report by the WHO focuses on efficiency as a key component of quality of care.[1] They advocate that the quality of care provided should be a product of its efficiency, given the resources of the region. Quality therefore is evaluated by how well regions use available resources to provide care. Utilization of this approach allows for variations between regions that may have vastly different access to resources, yet are providing healthcare to meet the needs of their immediate populations. If regions are using resources to full capacity, then they are deemed to be delivering efficient, high-quality care, regardless of how that care compares to other regions. Integration of this approach recognizes the efforts of remote or underdeveloped regions that may not have the infrastructure to provide substantial data on process and outcomes, but are able to demonstrate improved access to care, and utilization of all available resources in providing that care.

In conjunction with evaluations of efficiency, the WHO and other international organizations have proposed international health system benchmarks to promote ongoing evaluation and reporting of data to ensure healthcare quality. The WHO proposes five indicators to evaluate quality of care: (1) population health, (2) degree of health disparities/inequities, (3) health system responsiveness, (4) distribution of responsiveness, and (5) distribution of health system financial costs.[1] In contrast, other international groups propose more specific benchmarks with multiple indicators. The Healthcare Quality Indicators Project (HCQI) comprises 32 member countries and initially proposed 40 indicators to evaluate care quality: 8 indicators are specific to disease processes, 9 are related to health promotion, 2 center on mental health, 7 indicators reflect safety, and the remaining 10 indicators are specific for disease screening and prevention.[2] Revisions to these indicators has resulted in 62 indicators using 6 domains, which now include primary care, acute care, mental health, cancer care, patient safety, and patient experiences.[3] Similarly, the Dutch Healthcare Performance Report proposes 110 indicators that include quality, access, costs, and interconnected themes, such as efficiency, patient views, and degree of healthcare reform.[3] Lastly, the National Scorecard on United States Health System Performance advocates for 37 indicators centering on health outcomes, quality of care, access to healthcare, efficiency, and equity.[3] These international approaches to benchmarking continue to evolve, as monitoring and reporting of data on an international scale has inherent challenges.

Challenges with international data

Establishing proposed benchmarks is an important step toward ensuring healthcare quality on an international scale. However, operationalizing specific data to be gathered to meet each metric, as well as ensuring an infrastructure for ongoing monitoring, reporting, and evaluation of that data across countries is equally important. Key methodological considerations center on where the focus of the comparisons should lie—within structure, process, or outcomes data, as each component can vary substantially across health systems. Table 1 lists the advantages and disadvantages of reporting process measures compared to outcome measures. When determining what data to collect, it is important to acknowledge variations may exist on an international scale. The focus therefore should be on measures that are most meaningful for international comparisons. Key considerations should also include the replicability, reliability, and validity of the data abstraction

Table 1 International considerations of advantages and disadvantages of process vs outcome measures.

	Process measures	Outcome measures
Advantages	• Sensitive to change • Easily identify modifications to improve care delivery • Easy to measure and interpret • Identifies deficits rapidly • Smaller sample size required • Identifies patients-specific factors	• Promotes innovation and integration of new processes • Supports long-term health promotion interventions • Measures health system or population-level goals
Disadvantages	• Not broadly applicable across several diagnosis types; multiple measures may be needed for specific groups. • Easily outdated with technological advancements or organizational change • Easily manipulated and may skew accuracy of measurement	• Outcomes of interest may be delayed • Difficult to establish direct cause/effect between interventions and outcomes. • Data collection may be time consuming • Requires larger sample size

Table 2 Challenges for international data usage.

- ➢ Variation in available data
- ➢ Limited availability of data
- ➢ Different infrastructures and systems for recording data
- ➢ Poor data quality
- ➢ Different methods for data abstraction and definition of terms
- ➢ Determining optimal number of countries that must participate for data to be meaningful
- ➢ Lack of evidence to link specific indicators to outcomes across populations

method and source, along with considerations about advantages and disadvantages of process and outcomes measures.

To address these challenges, specific frameworks are proposed to evaluate quality of care on an international level, such as the Commonwealth Fund International Working group on Quality Indicators.[3] Key to these initiatives is tapping into existing national quality assessment programs to utilize existing datasets that allow for international comparisons. Many governmental organizations have established infrastructures for abstraction and storage of this data for international comparisons.[3] Specific quality assessment initiatives aimed at comparing data across international quality indicators often are already established by specific disease area.[3] Thus, international indicators using this approach are feasibly collected, used to generate reports and comparisons across countries, have some element of standardization, and provide important information on burden of diseases and healthcare costs.[4] Utilization of health information systems with electronic medical records may further aid in standardization of data reporting; however, variations among systems, access, and capabilities inevitably persist (Table 2).

International nursing benchmarks and use of data

Efforts toward international benchmarking and application of data among nurses have increased in recent years. While nurses have been involved in data acquisition and reporting for national or international research and quality initiatives for decades, recent efforts have focused on identifying indicators specifically applicable to nursing care and impact on outcomes across countries. Nurse-sensitive indicators are used to gauge quality of nursing care within health systems around the world. These indicators specifically link interventions of nursing care to patient and organizational outcomes, and are based on empirical evidence supporting these linkages. While nurse-sensitive indicators are well established within several specific countries, variations arise when examining these indicators on an international scale.

The use of a proposed international minimum data set for nursing quality indicators has been explored among nursing homes in the United States, Iceland, and Canada.[5] Common nurse-sensitive elements include urinary tract infections, dehydration, weight loss, bowel and bladder incontinence, frequency of urinary catheter use, methods for fecal impaction, use of tube feedings, prevalence of pressure ulcers, restraint use, and incidence of behavioral symptoms. Comparison of these standard measures across countries and health systems is an important initial step in establishing common international elements of quality in nursing.[5] Although challenges remain, specifically with reliability and validity of measures across health care delivery systems, work continues to establish common elements of nursing care quality.[6]

Nursing specialty organizations have also united on an international scale to identify potential nurse-sensitive quality indicators for specific patient populations. For example, the International Collaboration of Orthopedic Nurses (ICON), specifically created a hip fracture working group to evaluate evidence on best practices from nine countries, and use these data to propose international care standards across all settings.[7] The group created a 12-item tool of potential quality indicators, 11 of which are nurse sensitive. Indicators include early mobility, preventing malnutrition, catheter-associated urinary tract infection, management of pain, delirium assessment and prevention, prevention of pneumonia, prevention and management of constipation, prevention of venous thromboembolism, prevention of pressure injury, care transitions, and bone health. Timing of surgery is another quality indicator, but is not considered nurse sensitive. Evaluations of outcomes associated with these proposed indicators are ongoing.[7]

Other efforts toward establishing international nursing-sensitive indicators focus on expansion of existing measures within specific countries to determine applicability on a national scale. Nurse-sensitive indicators within Dutch hospitals include screening and observation of delirium, screening of malnutrition, prevention of pressure ulcers, and standardized pain assessment among postoperative, recovery room, and hospital patients.[8] Expansion of these indicators on an international scale may be a feasible approach toward uniform measurement across countries if applicable. Similarly, indicators used by the American Nurses Association are proposed, with the addition of components to capture patient satisfaction, ability of nurses to acknowledge and meet the needs of individual patients, and ensuring patient safety.[9] Regardless of the indicators used, efforts must include methods to ensure standardization of data abstraction and reporting, as well as specific considerations of each country's resources, existing infrastructures, regulatory oversights, and populations.

International resources for data and quality in healthcare

While the WHO is at the forefront of care delivery, data integration, and quality on an international scale, other organizations are instrumental in providing the necessary guidance and infrastructure to promote health outcomes. Organizations include the Pan American Health Organization, The Council of Europe and the European Region, and the Economic Co-operation and Development, as well as the World Organization of Family Doctors and the Cochrane Collaboration. Specifically within nursing, international groups include Sigma Theta Tau International, the International Nurses Association, as well as the International Council of Nurses.

The International Council of Nurses is the nursing arm of the WHO, and comprises 130 nursing organizations from around the globe that work collectively to advance the profession of nursing and positively influence healthcare policy. Key tenets of their work center on professional nursing practice, nursing regulation and education, and socioeconomic wellbeing of nurses throughout all global settings. Data utilization is at the core of their initiatives. Data are used by the group to inform advocacy efforts, specifically for policy change, as well as to establish guidelines for education, training, resource utilization, and establishment of evidence-based practice guidelines. Data are disseminated regularly through various publications by the group to advance their efforts on an international scale.

In addition to international nursing groups, there are several specialty nursing professional areas that have international representation geared toward care of specific patient populations. Examples include the World Federation of Neuroscience Nurses, International Association of Forensic Nurses, International Association of Multiple Sclerosis Nurses, and many others. Table 3 lists international nursing groups by specialty area.

Additional support and training for data utilization in nursing is available to developing countries. International resources include The Quality Assurance Project, The Joint Commission International, The International

Table 3 International nursing organizations.

➢ International Association of Forensic Nurses
➢ International Association of Nurse Anesthetists
➢ International Collaboration of Orthopedic Nurses
➢ International Childbirth Education Association
➢ International Child Health Nursing Alliance
➢ International Council of Nurses
➢ International Network for Doctoral Education in Nursing
➢ International Nurses Association
➢ International Organization of Multiple Sclerosis Nurses
➢ International Nurses Society on Addictions
➢ International Society of Nurses in Genetics
➢ International Society for Psychiatric-Mental Health Nurses
➢ International Transplant Nurses Society
➢ Therapeutic Touch International Association
➢ Sigma Theta Tau International

Society for Quality in Healthcare, and the Institute for Healthcare Improvement. These groups offer publications on key components of quality, as well as quality workshops at international conferences and in specific regions, and provide resources toward advancement of technological capabilities across nations to improve quality of care.

A multitude of resources are also offered within specific countries and may include local ministries of health, executive agencies, academic groups, and professional societies. Some countries have national societies for quality, as well as established benchmarks for specific outcomes, disease processes, and care delivery. While the standards and benchmarks may vary from being nonexistent to highly complex, an element of quality of care is inherent in all care delivered across all settings. Public, governmental, and private organizations continue important work to establish quality indices and benchmarks across nations, and to evaluate adherence and impact on outcomes and population health.

References

1. World Health Organization (WHO). *Quality and Accreditation in Health Care Services: A Global Review*. Switzerland: WHO; 2003.
2. Carinci F, Van Gool KV, Mainz J, et al. Towards actionable international comparisons of health system performance: expert revision of the OECD framework and quality indicators. *Int J Qual Health Care*. 2015;27(2):137–146.

3. Nolte E. *International Benchmarking of Healthcare Quality: A Review of the Literature*. Cambridge: RAND; 2010.
4. Hussey PS, Anderson GF, Osborn R, et al. How does the quality of care compare in five countries? *Health Aff.* 2004;23:89–99.
5. Jensdottir AB, Rantz M, Hjaltadottir I, Gudmundsdottir H, Rook M, Grando V. International comparison of quality indicators in United States, Icelandic and Canadian nursing facilities. *Int Nurs Rev.* 2003;.
6. Hjaltadottir I, Ekwall AK, Nyberg P, Hallberg IR. Quality of care in Icelandic nursing homes measured with minimum data set quality indicators: retrospective analysis of nursing home data over 7 years. *Int J Nurs Stud.* 2012;49(11):1342–1353.
7. MacDonald V, Maher AB, Mainz H, et al. Developing and testing an international audit of nursing quality indicators for older adults with fragility hip fracture. *Orthop Nurs.* 2018;37(2):115–121.
8. Stalpers D, Kieft RAM, van der Linden D, Kaljouw MJ, Schuurmans MJ. Concordance between nurse-reported quality of care and quality of care as publicly reported by nurse-sensitive indicators. *BMC Health Serv Res.* 2016;16:120–127.
9. Connolly D, Wright F. The nursing quality indicator framework tool. *Int J Health Care Qual Assur.* 2017;30(7):603–616.

Index

Note: Page numbers followed by *f* indicate figures and *t* indicate tables.